MARINE BIOLOGY

MEDITERRANEAN SEA

ECOSYSTEMS, ECONOMIC IMPORTANCE AND ENVIRONMENTAL THREATS

MARINE BIOLOGY

Additional books in this series can be found on Nova's website under the Series tab.

Additional E-books in this series can be found on Nova's website under the E-book tab.

OCEANOGRAPHY AND OCEAN ENGINEERING

Additional books in this series can be found on Nova's website under the Series tab.

Additional E-books in this series can be found on Nova's website under the E-book tab.

MARINE BIOLOGY

MEDITERRANEAN SEA

ECOSYSTEMS, ECONOMIC IMPORTANCE AND ENVIRONMENTAL THREATS

TERRENCE B. HUGHES
EDITOR

nova publishers
New York

Library of Congress Cataloging-in-Publication Data

ISBN: 978-1-62618-238-7

Published by Nova Science Publishers, Inc. † New York

Contents

Preface vii

Chapter 1 Ecosystem Services in the Mediterranean Sea: the Need for an Economic and Business Oriented Approach 1
Rafael Sardá

Chapter 2 Marine Pollution from Shipwrecks at the Sea Bottom: A Case Study from the Mediterranean Basin 35
Mario Sprovieri, Marco Barra, Marianna Del Core, Gabriella Di Martino, Luigi Giaramita, Serena Gherardi, Sara Innangi, Elvira Oliveri, Salvatore Passaro, Teresa Romeo, Paola Rumolo, Daniela Salvagio Manta, Stella Tamburrino, Renato Tonielli, Anna Traina, Giorgio Tranchida, Mattia Vallefuoco, Salvatore Mazzola and Franco Andaloro

Chapter 3 Disposal of Produced Formation Water from Offshore Gas Platforms in the Mediterranean Sea: A Parametric Study on Discharge Conditions Aimed at Mitigating Risks for the Marine Environment 65
Daniela Cianelli, Rossella Di Mento, Loredana Manfra, Federico Oteri, Anna Maria Cicero and Enrico Zambianchi

Chapter 4	Secchi Depth and Water Quality Control: Measurement of Sunlight Extinction *Javier Ibáñez Civera, Nicolás Laguarda Miró, Eduardo García Breijo and Rafael Masot Peris*	**91**
Chapter 5	An Environmental Monitoring Plan for the Construction and Operation of a Marine Terminal for Regasifying Liquefied Natural Gas (LNG) in the North Adriatic Sea *C. Virno Lamberti, M. Gabellini, C. Maggi, O. Nonnis, L. Manfra, S. Ceracchi, B. Trabucco, G. Moltedo, F. Onorati, G. Franceschini and R. Di Mento*	**115**
Chapter 6	Bioactive Peptides from the Venom of the Mediterranean Cone Snail *Conus Ventricosus* *Maria Carolina Spiezzia, Cristiano Chiarabelli, Maria Eugenia Schininà and Fabio Polticelli*	**135**
Index		**149**

PREFACE

In this book, the authors present topical research in the study of the ecosystems, economic importance and environmental threats to the Mediterranean Sea. Topics discussed include the need for an economic and business oriented approach to ecosystem services in the Mediterranean Sea; marine pollution from shipwrecks at the bottom of the Mediterranean Sea; the effects of produced formation water (PFW) on the marine environment of the Mediterranean Sea; Secchi depth and water quality control; an environmental modeling plan for the construction and operation of a marine terminal for regasifying liquefied natural gas in the North Adriatic Sea; and a study of bioactive peptides from the venom of the Mediterranean cone snail conus ventricosus.

Chapter 1 – The relationship of Mediterranean people with the sea still remains close after 4000 long years of history. Different empires and a large number of human generations have been using its environment in a granted way. All these uses have compromised the ability of the sea to operate correctly and put in danger some of its observed process, its structural units, and their functions and with them, the benefits that people get as ecosystem goods and services. Establishing connections between ecosystem change and people's benefit can lead to develop much more proactive approaches for conservation. The ecosystem service concept can be also useful at that point because it emphasizes the real notion of both protecting nature and benefiting man at the same time. Understanding of provision of ecosystem services (quantification), understanding of the benefits to human well-being from ecosystem services (valuation), and creating incentives for the sustainable provision of such ecosystem services (policies, good governance, alliances,...) should be recognized as a precondition for a sustainable future of our seas and coasts. In this way Innovative schemes for linking public and private efforts to protect ecosystems by ensuring the provision of ecosystem services need to be raised because whoever benefits from those services should have a responsibility in its proper care. Ecosystems provide many critical life support functions and benefits for human wellbeing and it is time to find the best way to ensure its good environmental status.

Chapter 2 – In order to properly adopt the international recommendations on the marine environment and biodiversity conservation, as well as fishing resources and human health, an increasing effort is incoming in the last two decades to study, prevent and mitigate the anthropogenic environmental impact in the Mediterranean sea. The main research efforts and policy strategies implemented in the last two decades are regarding oil pollution (dumping, spills, shipping, extraction, accidents, etc.), civil and industrial waste discharges and, recently, littering. Less known are the other pollution sources generally not visible and frequently distributed offshore. One of this cryptic sources of pollution is represented by military and commercial shipwrecks abandoned on the sea bottom. Actually, fuels, construction materials, army and transported goods deposited at the bottom of the sea can generate ecological damage thus modifying natural biological diversity and fishing activity with unpredictable hazards and dangerous effects on the marine ecosystem. In particular, indiscriminate and uncontrolled leakage of toxic material and organic/inorganic pollutants from those vessels could affect the wellbeing of the deep sea also for long-time.

Very few data are available on the effects of pollution produced by shipwrecks in the Mediterranean sea although thousand of military ships were abandoned at the seabed of the basin as effect of accidents during the first and the second word wars.

This contribution aims to explore the potential effects of contamination of a number of shipwrecks identified by multi-proxy geophysical tools at the bottom of one of the key areas of the Mediterranean sea, around the Sicily island, where intense ship traffic and fishing activities occur. Geophysical exploration and geochemical data (heavy metals, PAHs, PCBs, Pesticides, hydrocarbons, etc.) are here presented for selected cases and testify the high risk represented by this kind of point-source pollution on the marine system.

Chapter 3 – The offshore oil and gas production in the Mediterranean Sea has considerably grown over the last 20 years, introducing a wide range of potential risk factors for the health of marine ecosystems. Crucial impacts may derive from discharges of produced formation water (PFW), a by-product of gas extraction, which to date represents approximately 80% of the waste and residual yield from natural gas production activities. The disposal of PFW may be carried out through the reinjection into the reservoir, the transport onshore or the discharge into seawater. The authors focused their attention on PFW discharged by submerged outfalls in the Adriatic Sea (a shallow Mediterranean sub-basin characterised by an intense extraction and production activity).

The effects of PFW discharge on the marine environment depend on the characteristics of the effluent as well as on the dispersion processes occurring in the receiving ambient fluid. In order to assess these potential environmental impacts the authors apply a numerical dispersion model integrated with a monitoring plan developed for the Adriatic Sea.

The authors simulated the initial mixing of PFW discharged from a selected offshore gas platform during the summer season, and used field data (samples of ambient sea water, current intensity and direction measurements, temperature and conductivity profiles) collected in the vicinity of the platforms from 2008 to 2010 during monitoring surveys. The authors evaluated several different discharge scenarios (in terms of environmental conditions and effluent characteristics) and three working hypotheses for the discharge operations. The results of this parametric study allowed us to identify some sort of "optimal discharge practices" for PFW in the Adriatic Sea, which should promote the rapid dilution of the effluent and mitigate the impacts of the PFW on marine ecosystems.

Chapter 4 – Water is fundamental for life and its quality control is becoming more and more important. In fact, there are vast strategic actions and laws trying to manage it. An example is the European Water Framework Directive 2000/60 EC that summarizes and manages all the actions to water quality control in European Water bodies. In this sense, Turbidity is one of the water parameters that may present some problems when being measured. This is due to the fact that the Secchi Disk measure is a worldwide accepted procedure but it may have some limitations such as the subjectivity of the measure (depends on the ability of the operator), its dependence on environmental factors and its limitation in shallow water bodies where it may reach the bottom without having disappeared.

The goal of this chapter is to collaborate in the improvement of turbidity measurement by suggesting an alternative procedure to Secchi Depth determination based on a new device called LUZEX. This new device uses optoelectronic measurements of sunlight extinction with depth in water bodies to calculate a sunlight extinction coefficient for different wavelength bands.

Several measurements have been done in continental, sea and transitional water bodies in order to carry out a comparison between the measured values for Secchi Disk and the obtained values for Secchi depth based on a mathematical model and the data given by the authors' LUZEX device.

As a conclusion, the suggested method to substitute Secchi Disk measurement is quite good as the authors have obtained regression coefficients (R^2) up to 0.9967 in the authors' best models.

Chapter 5 – Natural gas is mainly used for generating electricity and for residential uses. Italy does not extract a sufficient amount of natural gas to meet domestic consumption. Given the growing demand for natural gas and the need to differentiate supply routes to make Italy less dependent on individual countries, in 1998 the project of the "Adriatic LNG Terminal" was authorized. The project consists of the construction and operation of an offshore terminal for regasifying Liquefied Natural Gas (LNG) and its pipeline for transferring it to land, in the Northern Adriatic Sea (Porto Viro, Rovigo). The Italian Ministry of the Environment has assessed the environmental compatibility of this project prescribing monitoring to verify possible impacts on the marine environment associated with the construction and operation of the structure. The Institute for Environmental Protection and Research (ISPRA, formerly ICRAM) elaborated and carried out the monitoring plan according to environmental prescriptions. The monitoring involves environmental studies in the marine area (where the terminal and part of the pipeline is located) and in the wetland area consisting of a lagoon and a fish valley, crossed by the connecting pipeline before reaching the land. Three monitoring phases have been put on: 1) previous to the construction of the structures, 2) during the construction and 3) during the terminal's operation.

This paper describes the Adriatic LNG Terminal and its pipeline Monitoring Plan (ALNGM). This plan provides the application of an integrated approach in order to analyze the possible environmental alterations that could occur.

Chapter 6 – Cone snails are marine predator gastropods that represent an incredible source of bioactive peptides also known as conopeptides. Their venom is a complex mixture of different types of (macro)molecules but in all the species the most abundant component is represented by conotoxins, disulphide-rich peptides that display the cysteine knot structural motif. Because of their high specificity and potency to bind and inhibit ion channels and receptors, conotoxins represent powerful tools for neuroscience as well as lead compounds for drug discovery. *Conus ventricosus*, a worm-hunting cone snail, is the single cone snail species living in the Mediterranean Sea and the characterization of its venom started about ten years ago. The first *C. ventricosus* peptide isolated and characterized was Contryphan-Vn, a 9 amino acids peptide which displays activity on both invertebrate and vertebrate Ca^{2+} dependent K^{+} channels. Later, *C. ventricosus* venom peptides profiling led to the discovery of an additional conopeptide of the Leu-Contryphan family, whose molecular target is still unknown, and two novel conotoxins, named α-Conotoxin-Vn and Conotoxin-Vn2. The former one targets nicotinic acetylcholine receptors and the latter most probably voltage-gated Ca^{2+} channels. Recently a recombinant expression system has been developed to increase the yield in soluble Conotoxin-Vn2 and perform structural and functional analysis. This analysis revealed that conotoxin-Vn2 displays a high toxic activity towards insects larvae, providing new means for the development of a novel generation of environmental friendly pesticides.

In: Mediterranean Sea
Editor: Terrence B. Hughes

ISBN: 978-1-62618-238-7

Chapter 1

ECOSYSTEM SERVICES IN THE MEDITERRANEAN SEA: THE NEED FOR AN ECONOMIC AND BUSINESS ORIENTED APPROACH

Rafael Sardá*
Centre d'Estudis Avançats de Blanes (CEAB-CSIC), Girona, Spain

ABSTRACT

The relationship of Mediterranean people with the sea still remains close after 4000 long years of history. Different empires and a large number of human generations have been using its environment in a granted way. All these uses have compromised the ability of the sea to operate correctly and put in danger some of its observed process, its structural units, and their functions and with them, the benefits that people get as ecosystem goods and services. Establishing connections between ecosystem change and people's benefit can lead to develop much more proactive approaches for conservation. The ecosystem service concept can be also useful at that point because it emphasizes the real notion of both protecting nature and benefiting man at the same time. Understanding of provision of ecosystem services (quantification), understanding of the benefits to human well-being from ecosystem services (valuation), and creating incentives for the sustainable provision

* Tel +34 972 336101, Fax +34 972 337806, Email: sarda@ceab.csic.es.

of such ecosystem services (policies, good governance, alliances,...) should be recognized as a precondition for a sustainable future of our seas and coasts. In this way Innovative schemes for linking public and private efforts to protect ecosystems by ensuring the provision of ecosystem services need to be raised because whoever benefits from those services should have a responsibility in its proper care. Ecosystems provide many critical life support functions and benefits for human wellbeing and it is time to find the best way to ensure its good environmental status.

Keywords: Ecosystem services. Valuation. Mediterranean Sea

INTRODUCTION

For at least 4,000 years, since the Egyptians, Minoans and Phoenicians were trading across and, probably even early by other communities fishing in their waters, people have been using the Mediterranean Sea. The Mediterranean region has been traditionally recognized as a crossroad of marine routes, biota and civilization. This relatively calm and beautiful enclosed sea with numerous islands as magnificent posts, have provided an ideal setting for humans around their waters using their multiple components mostly in a granted way. Fisheries, navigation, battles, and commercial trading were for more than 3.000 years the main human uses of this environment that constituted the central sea of western civilization and the trade link to the eastern world. The western migration of people to America shifted the center of gravity to the Atlantic coast of Europe and the importance of the Mediterranean decreased. However, the opening of the Suez Canal in 1869 provided an easy route to Asia and increased again the importance of the Mediterranean for worldwide economic activities and human development; a non-ending history up until today.

In 2012, the 21 Mediterranean countries had around 7% of the entire world population (510 million people; CIA World Factbook). Although slowing down its growth, population is still increasing in the Mediterranean region. Some of the trends connected with this growth such as an enormous increase of economic activities in the region, with especial attention to international tourism arrivals and coastal development in the northern rim, a growing population in urban environments especially in the southern rim (it is expected that almost 75% of the total population will inhabit urban areas larger of 10.000 inhabitants by 2025), and an exploding aging population, accelerated new demands on modern facilities and infrastructures, altogether incrementing the pressure on the natural systems of its environment. Due to such trends, during the last decades we have seen a cascade of cumulative pressures on its ecosystems components that yield a tremendous transformation of the Mediterranean natural environment.

The drivers of change seen for the Mediterranean mirror a global anthropogenic-driven transformational change (Crutzen, 2002; Steffen *et al.*, 2007; Rockström et al., 2009). In recognition to such transformation, a decade ago, United Nations launched the Millennium Ecosystem Assessment report (MEA, 2005). This report constituted a global audit on world's ecosystems providing summaries and guidelines for decision-makers. The report concluded that ecosystems have declined more rapidly and extensively over the past 50 years that at any other comparable time in human history, and advised on the issue that this degradation jeopardize not only ecosystems around the globe but also human activities. The report established the link between human activities and good environmental status because healthy ecosystems provide natural conditions that are indispensable for human well-being and societal welfare.

The new ideas included in the MEA report (highly dependence of human populations on natural goods and ecosystem services, indivisible link nature-man inside social-ecological systems, ecosystem approach for management, ...) scaled up during the past decade to be settled into the international environmental policy agenda. Between all these applications, in Europe, the new Marine Environmental Policy (Marine Strategy Framework Directive (MSFD) and the Integrated Maritime Policy (IMP)) was conceived to achieve a Good Environmental Status (GEnS) and a sustainable use of marine ecosystems. This new policy structured a common vision and a holistic, integrated approach to the management of the marine environment using the Ecosystem Approach as its basic frame of reference (Farmer et al., 2012). The new Marine Environmental Policy emphasizes the necessity to protect, maintain and/or enhance natural resources, both natural goods and ecosystem services.

The blue waters of the Mediterranean, as well as other enclosed seas and oceans elsewhere, contain myriads of cells that use sunlight to produce organic matter through the photosynthesis process. This organic matter travels around complex trophic chains giving us as a humans living support. This produced biodiversity also help us to obtain other types of goods but also to support other fundamental ecosystem functions such as water depuration, coastal protection or carbon fixation, between others, that benefit man. During the last three decades, an effort has been made to bring not only these ecosystem functions into the concept of ecosystem services through the recognition of the benefits they give to man (Daily, 1997; de Groot, 2002; Gómez-Baggethum *et al.*, 2010), but also to value them in monetary terms, and recently to incorporate this jargon into markets and payment schemes.

Ecosystem services help to make the Mediterranean region much more habitable but despite this fact still they are not entirely understand and widely appreciated by people.The ecosystem services provided by the Mediterranean Sea are today increasingly compromised by human activities. Regional assessments done for the Mediterranean confirm that the capacity of our sea to continue in the delivering of ecosystem goods and services is declining as a result of human activities (Coll *et al.*, 2010, 2012; Sala *et al.*, 2012) following global trends observed elsewhere (MEA, 2005; Worm *et al.* 2006; Halpern *et al.* 2008; Ewing *et al.*, 2010). The new European marine policy was also aimed to stop and reverse this degradation. Having all the above considerations in mind, the aim of this article is twofold; on one hand, to review fundamental ideas around the ecosystem services concept and to analyze how human-related stressors translate into changes not only of the present people's welfare but also of future generation's welfare, and, on the other hand, to advocate for new public-private partnerships to raise innovative models of collaboration that help in the maintenance and/or restoration of vital ecosystem services. Establishing connections between ecosystem change and people's benefit can lead to develop a much more proactive approach in the management of the Mediterranean Sea natural resources.

Concepts on Ecosystem Services and Its Classification

An ecosystem consists of a biological community (biotic component) together with its abiotic environment, interacting as a whole (Chapin *et al.*, 2002). Ecosystems are complex (structurally and functionally) adaptive systems that are controlled both by internal and external factors providing a variety of good and services upon which people depend. Ecosystem goods and services are delivered through various combinations of ecosystem functions which are in turn delivered by different components of its biodiversity.

The notion that ecosystems support human societies and that natural resources should not be seen as discrete entities is not new and a variety of terms were offered in the past to describe this issue. It was firstly mentioned in the report "Study of Critical Environmental Problems (SCEP, 1970; reviewed by Daily, 1997) as "environmental services", services that would decline if the functions of ecosystems also decline. The concept was slightly modified by Ehrlich *et al.* (1977) "public services of the global ecosystems", by Westman (1977) "nature's services", and finally it became named as "ecosystem services" (Ehrlich and Ehrlich, 1981; de Groot 1992; Costanza *et al.* 1997; Daily 1997; Gilbert and Janssen 1998; de Groot *et al.*, 2002; Boyd and Banzhaf 2007). In 1997 Daily proposed the following definition "*conditions and processes through which natural ecosystems, and species that make them up, sustain and fulfill human life*". The concept was lately popularized and formalized in the Millennium Ecosystem Assessment (MEA, 2005) in a much simple way "*the benefits that people obtain from ecosystems*". Both definitions established the linkages between ecosystem services and human well-being by recognizing that ecosystems, if sustainably managed and protected, benefit current and future people and societies, bringing the idea that the concept of ecosystem services shows the flow of benefits from nature to people, providing a framework that can be used in the management of public goods.

In today's world, natural resources (ecosystem goods and services) cannot be treated as discrete entities that need to be analyzed separately; they are dependent of the social and economic systems with which they interact. The concept of social-ecological systems (Berkes and Folke, 1998) has been developed recently to analyze this complexity. Social-ecological systems (Figure 1) are "*complex adaptive systems in which humans are part of nature and the dynamics of both dimensions are strongly linked at equal weight*". These coupled, co-evolving models, by definition should focus on the ability of the management system to respond to feedbacks from the environment having deeply into consideration the tendencies of goods and services we obtain as benefits from the environment.

Social-ecological systems can be analyzed through different information platforms. Recently Cooper (2012) have proposed the Driver-Pressure-State-Welfare-Response (DPSWR) framework (an evolution of the DPSIR framework, EEA, 1999) to adequately organize information of the interrelation between the human and the natural sub-systems inside. Human systems (people's capabilities and their activities) become drivers of change (D). They pressure constantly or in pulses natural related systems (P). Those Natural systems (structural units and the functions they made) alter their status (S) that in turn can translate into the degradation of fundamental natural resources used by man (natural goods and ecosystem services) diminishing human welfare (W).

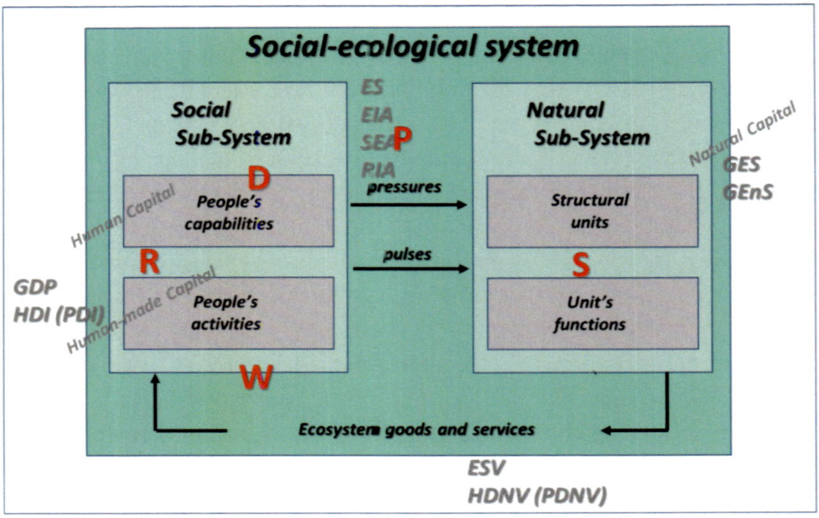

Figure 1. Schematic representation of a social-ecological system with accounting frameworks for its individual parts.

The recognition of such degradation should allow man to made adequate policy responses (R) to solve the pattern of accelerated degradation. The information generated through the DPSWR framework expressed issues in a highly inter-related form which is something that we cannot observe if we use other sets of indicators that inform different pieces of the social-ecological system puzzle in isolation. Human systems are normally monitored through a myriad of economic indicators such as Gross Domestic Product (GDP) or Household and/or Personal Disposable Income (HDI, PDI). When man pressures are analyzed, a large set of impact indicators commonly used in Environmental, Strategic and/or Regulatory Impact assessments (EIA, SEA, RIA) are developed, as well as environmental standards (ES) are set up to have these pressures under control. Indicators for Natural systems (besides the ones used in the Academia) are much more recent; In Europe, the Marine Strategy Framework Directive has introduced recently the notion of Good Environmental Status (GEnS, following the nomenclature recommended by Borja et al. (2010)) in a similar way that the Water Framework Directive did with the Good Ecological Status (GES) indicators; GEnS works with a set of multiple indicators following 11 environmental descriptor. Finally, the benefits obtained by man from natural systems can be assessed using market indicators for marketed natural goods or assessing ecosystem service valuation (ESV) accounting also for Household and/or Personal Disposable Natural Value (HDNV, PDNV).

Different classifications to hierarchically accommodate ecosystem services can be found in the literature (de Groot, 2002; Farber *et al.*, 2006; MEA, 2005; TEEB, 2010). These classifications relate ecosystem services to a function analysis approach (de Groot, 2006) because in order to achieve the level of ecosystem services provision by society we need to identify what functions determine what services.

The list below address the four categories of ecosystem services following MEA (2005) including the changes added by TEEB (2010); provisioning services, regulation services, supporting services (including habitat services) and cultural services.

- *Provisioning services:* cover materials or energetic outputs from ecosystems.- Food, Water, Raw Materials, Medicinal resources, Ornamental resources, and Genetic resources.
- *Regulating services:* cover benefits obtained from the regulation of ecosystem processes.- Climate regulation, Carbon sequestration, Disease regulation and Biological control, Water flow regulation and purification. Air quality management. Maintenance of soil fertility, Erosion prevention, and Pollination.
- *Cultural services:* cover non-material (intellectual/cognitive /symbolic) uses from ecosystems.- Spiritual and religious, Recreational and Tourism, Aesthetic enjoyment, Inspirational for culture, art and design, Educational and Cognitive development, and Cultural Heritage.
- *Supporting services:* necessary for the production of all other ecosystem services.- Soil formation, Nutrient cycling, Primary production, Nursery services, Gene pool protection, and Habitat maintenance.

Ecosystem services may not have to be utilized directly and can be differentiated as intermediate a final services (fish production is a final service, primary production is an intermediate service, necessary to get the final service). This separation into intermediate and final services, with the latter yielding welfare benefits is important in any possible valuation exercise because it helps to reduce the risk of double counting the welfare benefits provided to humans by often complex ecosystem functioning. Using another example that will be employed lately, in coastal areas, geodynamics and sediment transport can be considered an intermediate service (provisioning, regulation) of a final service consistent in the creation and maintenance of beaches, dunes and other places for humans which can be seen as benefits for people in flood/storm buffering, shoreline stabilization or recreational activities (security, basic mater for good life, good social relations,…).

Another important consideration about ecosystem services is the need to translate them into benefits that can improve human well-being (Fisher and Turner, 2008). Ecosystem services constitute the link between ecosystems, the way they operate and those things that humans benefit from, in other words, ecosystem services are the ecological phenomena, and the benefit is the realization of the direct impact on human welfare, for instance, using the above commented example, the production of fish is a service, the income derived from fisheries is a benefit. The fundamental concept behind is that ecosystem services are phenomena that provide benefits for man.

Ecosystem Services Valuation

Valuation Techniques

Valuation is the process of assessing the contribution of a particular object or action to meeting a particular goal, whether or not that contribution is fully perceived by the individual. One reason for the persistent under-valuation of coastal and marine ecosystems is that its economic value has been based on a very narrow definition of benefits for man. Many people tended to see the value of natural ecosystems only in terms of the raw materials and physical products that can generate for our human production and consumption (especially focusing on commercial activities and profits, those values tan can be marketed). These direct uses however represent only a small proportion of the total value of coastal ecosystems, which generate economic benefits far in excess of just physical products or marketed commodities. Confining concepts of ecosystem value to these benefits alone would constitute a huge underestimation, and covers only the tip of the total value.

The most appropriate framework to assess the overall economic value of a particular ecosystem is that of Total Economic Value (TEV). As we have seen previously, ecosystems give rise to a large range of natural goods and ecosystem services that include direct use values, indirect use values and passive (non-use)-use values, the latest comprising option, existence and bequest values (Costanza and Folke, 1997).

- *Direct-use values* are determined by the contribution an environmental asset makes to current production or consumption through direct use of the site.
- *Indirect-use values* include benefits derived from ecosystem services provided to support current production and consumption.
- *Option values* are the premium that consumers are willing to pay for a utilized asset; i.e., to avoid the risk of not-having it available in the future.
- *Existence values* are obtained due to satisfaction of merely knowing that the asset exists, even if there is no intention to using it.
- *Bequest values* are conceptualized when the above satisfaction comes from the idea of leaving the asset for further generations.

Although the goal of ecosystem services valuation is efficient allocation (*i.e.* to allocate scarce ecosystem services among competing uses such as development and conservation), other goals can be identified (Daly 1992): (a) assessing and insuring that the scale or magnitude of human activities within the biosphere are ecologically sustainable; (b) distributing resources and property rights fairly, both within the current generation of humans and between this and future generations, and also between humans and other species; or (c) efficiently allocating resources as constrained and defined by a) and b) above, and including both market and non-market resources, especially ecosystem services. Because of these multiple goals, valuation must be conducted from multiple perspectives, using multiple techniques (including both subjective and objective methodologies) (Brenner, 2007).

Many natural goods and ecosystem services are not traded. In most of the cases undervalued by the market, subjected to prices which are highly distorted, or that have characteristics of public goods which mean that they are not adequately allocated or priced. For these reasons, their value cannot be expressed accurately via market prices. To solve this problem, a range of different valuation techniques have been identified to do the job. There are different methodologies available to estimate the economic value of natural goods and ecosystem services (see list below), however each valuation methodology has its own limitations, often limiting its use to a select range of ecosystem services, and still, many of them are difficult to compare in between and yield different results for the same ecosystems service under valuation.

Market price based values

- *Market values;* value based on market prices taking into account government interventions such as taxes and subsidies.
- *Surrogate price values;* the value of a non-marketed asset is based on the market value of an alternative product providing similar benefits.
- *Productivity change values;* the value is based on the change in quality and/or quantity of a marketed good and the associated change in the total net market value.
- *Damage cost avoided;* the value of the asset is equal to the value of the economic activity that it protects.
- *Expected values;* value based on potential revenues multiplied by its probability of occurrence.

Cost based values

- *Replacement cost;* value based on the price of the cheapest alternative way of obtaining that service.

Revealed preference values

- *Travel cost method;* value inferred from the cost of travel to a site and its expenses.
- *Hedonic price values;* the value is based by computing its individual components which can be determined through regression analysis.

Stated preference values

- *Contingent Valuation;* the value is obtained by asking through questionnaire survey techniques how much people is willing to pay to prevent loss of, or enhance an ecosystem good or service.
- *Choice experiments;* the value is obtained similarly to Contingent Valuation involves also asking respondents to select their preferred package of environmental goods at different prices and then inferring specific components values via econometric analysis.
- *Transfer of values;* the value is estimated by transfer other obtained values from another similar context or location to allow us to estimate the one we desire.

Obviously the choice of any valuation technique will depend on the availability of resources for the study. Normally, the full suite of ecosystem valuation techniques will be required to account for the economic value of goods and services provided by a particular ecosystem. Table 1 is showing a list of ecosystem services together with its normal techniques used for valuation and the evaluation for a further transferability of the value across sites following Farber *et al.* (2006).

Ecosystem Services Valuation in the Mediterranean

We have seen a complete range of possibilities to make valuation on ecosystem services depending on the spatial scale used (ecosystems explored) and the employed methodologies (valuation techniques). A huge compilation effort has been made recently by the Marine Ecosystem Services Partnership to construct a database (http://www.marineecosystemservices.org/explore) of peer and non-peer reviewed empirical studies to extract ecosystem service valuation data. Table 2 provides information on ecosystem services valuation exercises from the Mediterranean region mostly extracted from this database. Information has been assessed based on ecosystem service type, land-cover type and geographical area of the Mediterranean, as well as on valuation methods. From the data presented we can assess that still we are in the infancy of this kind of approximations, we have very few empirical data.

Table 1. Categories of ecosystem services, economic methods for valuation and transferability across sites (adapted from Farber et al., 2006)

Ecosystem services		Amenability to Economic valuation	Most appropriate method for valuation	Transferability across sites
Provisioning	Water supply	High	AC,RC,M,TC	Medium
	Food	High	M,P	High
	Raw materials	High	M,P	High
	Genetic resources	Low	M,AC	Low
	Medicinal resources	High	AC,RC,P	High
	Ornamental resources	High	AC,RC,H	Medium
Regulating	Gas regulation	Medium	CV,AC,RC	High
	Climate regulation	Low	CV	High
	Disturbance regulation	High	AC	Medium
	Biological regulation	Medium	AC,P	High
	Water regulation	High	M,AC,RC,H,P,CV	Medium
	Soil retention	Medium	AC,RC,H	Medium
	Waste regulation	High	RC,AC,CV	Medium to high
	Nutrient regulation	Medium	AC,CV	Medium
Cultural	Recreation	High	TC,CV	Low
	Aesthetics	High	H,CV,TC	Low
	Science & education	Low	Ranking	High
	Spiritual & historic	Low	CV	Low
Supporting	Soil formation	Medium	CV,AC,RC	Medium
	Habitat maintenance	High	CV,TC	Medium
	Nursery service	High	CV, TC	Medium

Most appropriate method for valuation: AC = avoided cost; CV = contingent valuation; H = hedonic pricing; M = market pricing; P = production approach;
RC = replacement cost; TC = travel cost.

Table 2. Ecosystem service valuation data from the Mediterranean region (adapted from the Marine Ecosystem Services Partnership organization).(data presented in black boxes were obtained by contingent valuation techniques; data presented in dark grey obtained by travel cost methodology; data presented in light grey obtained by transfer value methodology and the rest of data obtained by other methodologies)

		SPAIN	ITALY	GREECE	TURKEY
Continental Shelf	Atmospheric gas & Climate regulation				
	Disturbance regulation				
	Freshwater regulation & Supply	1287 (1)			
	Erosion control & Soil formation				
	Nutrient regulation & Cycling	1787 (1)			
	Waste treatment				
	Pollination				
	Biological control	49 (1)			
	Habitat refugium			366000-1708000 (10)	
				6.2 (11)	
	Aesthetic & Recreation				
	Cultural & Spiritual	86 (1)			
Beach or Dune	Atmospheric gas & Climate regulation				
	Disturbance regulation	67400 (1)			
	Freshwater regulation & Supply				
	Erosion control & Soil formation				
	Nutrient regulation & Cycling				
	Waste treatment				
	Pollination				
	Biological control				
	Habitat refugium				
	Genetic resources				
	Aesthetic & Recreation	36687 (1)	13.3-28.4 (3)		11 (13)
	Cultural & Spiritual	59 (1)			
Seagrass bed	Atmospheric gas & Climate regulation				
	Disturbance regulation				
	Freshwater regulation & Supply				
	Erosion control & Soil formation				
	Nutrient regulation & Cycling	24228 (1)			
	Pollination				
	Biological control				
	Habitat refugium				
	Genetic resources				
	Aesthetic & Recreation				
	Cultural & Spiritual				

Salty wetlands	Atmospheric gas & Climate regulation				
	Disturbance regulation	766 (1)		42.5 (8)	
	Freshwater regulation & Supply			43.3 (8)	
	Erosion control & Soil formation			40.9 (8)	
	Nutrient regulation & Cycling			44.4 (8)	
	Waste treatment	13376 (1)			
	Pollination				
	Biological control				
	Habitat refugium	497 (1)	30.4 (4)	40.2 (8)	
			41-65 (5)		
				12.0-50.3 (9)	
	Genetic resources				5064.7 (12)
	Aesthetic & Recreation	64 (1)			
	Cultural & Spiritual	445 (1)	1300 (3)		
	Fisheries		2607 (6)		
	Multiple		607-660 (7)		

(1)	per ha ($)	Brenner J., Jimenez, J., Sardá, R., Garola A. (2010).
(2)	per ha (euro)	Ariza, E. Ballester, R., Rigall, R., Saló, A., Roca, E., Villares, M., Jimenez, J.A., Sardá, R. (2012)
(3)	per ha ($)	Gren, I.M., Soderqvist, T. (1994).
(4)	per ha ($)	Nunes, P., Rossetto, L. (2004).
(5)	per household (French Francs)	lberini, A., Rosato, P. (2004).
(6)	per household (French Francs)	Zanatta, V., Alberini, A., Rosato, P, Longo, A. (2005).
(7)	per fisherman ($)	Nunes, P., Silvestri, S. (2008).
(8)	per person (euro)	Ragkos, A., Psychoudakis, A. (2006).
(9)	per person (euro)	Stithous, M. (2009).
(10)	per year	Togridou, A., Hovardas, T., Pantis, J.D. (2006).
(11)	per person (euro)	Peters, H., Hawkins. J. (2009).
(12)	per ha year($)	Gurluk, S., Rehber, E. (2006).
(13)	per person (UK and)	Blakemore, F., Williams, A. (2008).

In addition, the information provided normally is difficult to compare by the different techniques used and the different values expressed (value per area-year, value per person, value per household, ..); nevertheless, it can serve to understand the urgent need to carry out new empirical studies that need to be commissioned to end up with more satisfactory estimations.

Using this database we can see some of its examples. This section presents three case studies on ecosystem service valuation for the Catalan coast (North-Western Mediterranean) and its open waters in which we have been involved. These examples vary on the geographical scale that is considered for the analysis (from large spatial areas as the entire Catalan coast to a particular beach environment in the municipality of Lloret de Mar) and can serve to understand the way in which ecological and socioeconomic assessments can be carried out and can be used by policy makers or managers to find out the best policy decisions.

Ecosystem services valuation in the Catalan coast

Catalonia (32,105 km^2) is located at the North-eastern Spanish Mediterranean coast (Figure 2). Seven per cent of its surface is occupied by 70 municipalities that are in its coastal fringe (699 km long of which 270 km are beaches) and that form part of 12 coastal counties. These coastal environments (coastal counties, called "*comarques*" in Catalonia) are affected by different socio-economic activities, being most relevant industrial and urban development, services (mainly tourism) and agriculture. The Catalan coast natural environment is composed of a diverse mixture of forests, grasslands, wetlands, rivers, beaches, seagrass beds, and platform environments that provide many different valuable goods and ecosystem services to human beings. A valuation exercise (Brenner, 2007; Brenner *et al.*, 2006, 2010) was developed using a value transfer methodology to assess the ecosystem services annual flow that coastal ecosystems were providing.

Ecosystem services valuation provided by the Catalan coast was computed using the method proposed by Troy and Wilson (2006) that follows spatial value transfer analysis. The study focused on terrestrial and marine goods and services which are not counted in the economic markets, therefore marketed goods, such as commercial offshore fisheries (> 50 m depth), aquaculture and agriculture were not part of the scope of this study. Using this method, the analysis of the ecosystem services value (ESV) annual flow by relevant coastal management units was then computed. Ecosystem services classification followed the scheme of Farber *et al.* 2006.

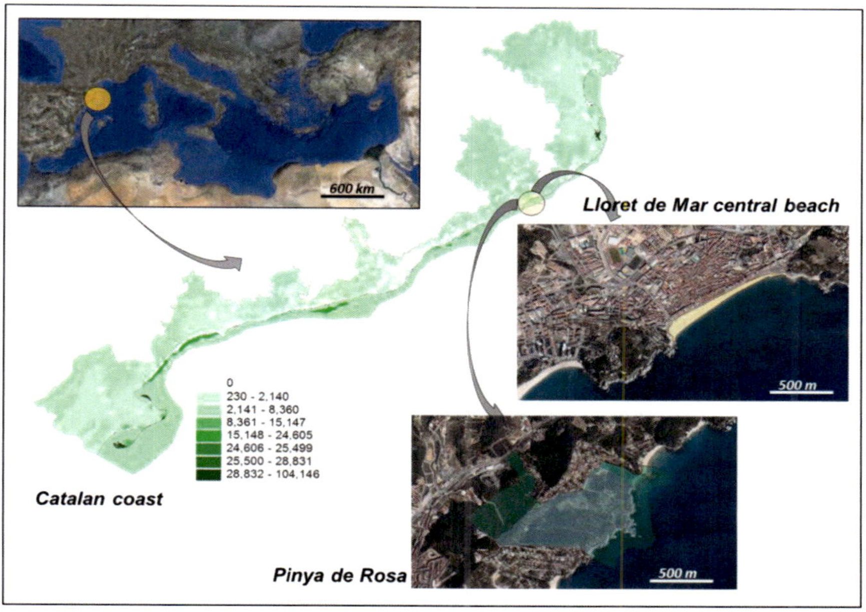

Figure 2. Cartographic representation of examples given for ecosystem service valuation in the Catalan coast. Ecosystem services value included for the Catalan coast is given in 2004 US$ per ha and year.

The geographical definition of the Catalan coastal zone was made by its coastal counties ("comarques") and near-shore marine areas (less or equal to 50 m depth). The terrestrial and marine cover typology used in this study constituted a merge of the Catalonia habitats (DMAH, 2006), the Catalan Sea bathymetry (DARP, 2000) and the seagrass beds vector layers (DARP, 2002). For the transfer value approach, scientific literature included in the analysis was basically empirical analyses in peer reviewed journals and book chapters that use conventional economic valuation methods and restricted to preference based valuations (*e.g.,* travel cost, hedonic pricing, contingent valuation); only meta-analysis of peer reviewed and non-peer reviewed studies were used when no data was available on the other group. Studies were limited to results that can readily be translated into spatial equivalencies focused on similar socioeconomic and biophysical regions as the North western Mediterranean, and primarily on non-consumptive resource use and ecosystem services (*i.e.,* non-market value). Estimation of ecosystem services value per area unit were standardized to average 2004 US dollars (USD) equivalents per hectare and per year to provide a consistent basis for comparison (for further comparisons, we used 1 USD = 133.94 Peseta, and 166.38 Peseta = 1 Euro set in 1994 by the Bank of Spain). The final calculation of the value of annual flow followed the recommendations of Bateman *et al.* (2002) on the application of Geographical Information System to environmental economics. An area of 931,460 ha was valuated in this study, 22.2% of this area corresponded to the coastal and marine domain (saltwater wetlands, beach and dunes, seagrass beds and the platform shelf shallower than 50 m depth) while 77.8% was terrestrial. The annual flow of non-market value of ecosystem services shown for each land and marine cover type in the Catalan coast (USD/yr in 2004) was calculated as 3.2 billion dollars (59% of this value coming from the terrestrial environment and 40.3% for the marine one). The value estimates calculated in the present study were compared with current market economic indicators to explore the contribution made from the natural capital in the coastal zone of Catalonia. Results show that total ESV flow was comparable to be around 2.8% of the Gross Domestic Product (GDP) in the study area ($114.8 billion USD in 2004), and the total ESV flow was around 4.3% of the available household disposable income ($74.4 billion USD in 2004).

The Natural area of Pinya de Rosa (Girona, Catalonia-Spain)

Pinya de Rosa is a beautiful natural coastal area located between the municipalities of Blanes and Lloret de Mar (Girona, Catalonia-Spain) (Figure-2). Despite the fact that the region form part of a highly touristic region (one of the most visited sites in Spain) the area have been preserved due to its private ownership for more than 60 years (all the period of high coastal development in Spain). After the owner's death, Pinya de Rosa was declared for sale at a price of 12 million euros in 2002. A socio-environmental public platform was born at that time to protect the space from a radical transformation by developers into an urbanized area. The platform was successful enough and finally, in 2003, Pinya de Rosa was formally protected by the Law 25/2003 of the Autonomous Government of Catalonia under one of the most important protected figures in Catalonia ("*Paratge Natural d'Interès Nacional*"-PENIN). Ninety-six ha of land were preserved under this Law.

Before protection, a scientific assessment was carried out. From all the information provided (Sardá et al., 2002), an ecosystem services valuation was established for this particular area to guide a social cost-benefit analysis before decision was made. The transfer value assessment methodology used in the previous example for the Catalan coast valuation was also used here. The 96 ha protected gave an ESV flow of around 400,000 euros per year. Although market and non-market valuations are difficult to compared, the provision value of the ecosystem services in Pinya de Rosa could be balance against investment values to be spend in land transformation and further societal benefits.

After protection, in 2004, the Government of Catalonia made an offer to heirs worth about 4 to 5 million euros for the acquisition of Pinya de Rosa as non-urbanized protected area.

The offer was rejected, and, at the same time, the property owners demanded 24 million euros to the Government for compensation for the land-use reclassification done to protect this area. Some years later, in 2008, a private investment company group bought 70 ha of Pinya de Rosa property for 24 million euros, a higher quantity to the one initially demanded at the beginning but enough to compensate the owners. The intention behind this acquisition was, and it is still, obscure because no transformation can be done in an area protected by an strong Law in the future but it shows how natural pristine areas can sometimes been undervalued by people, even for market prices.

The central beach of Lloret de Mar (Girona, Catalonia-Spain)

In Spain, as in many regions of the Mediterranean, beaches play a key role in the maintenance of the Tourism Industry, an essential sector for the economic welfare of some countries (Sardá and Fluvià, 1999; Sardá, 2001). Yepes (2004) described how the 0,001% of the Spanish surface (beaches that holds the "sun and beach" tourism model), are indirectly responsible of more than 10% of the Spanish Gross Domestic Product.

Consequently, beaches are considered to be one of the country's major assets. In the case of Lloret de Mar, its central beach (1.3 km; 5.6 ha) is considered one the most popular beaches in the entire Catalonia. Sardá *et al.* (2009) estimated in 22,036 the users per day of this beach during high season (the most visited days during summer, around 30 days during July and August).

As other coastal ecosystems beaches play multiple functions, being three the most important ones: to act as natural reservoirs, to offer coastal protection, and to provide human recreation.

A long list of ecological services is provided by these three assigned functions. The recreational service of the Lloret de Mar central beach was assessed using in this case the Travel Cost Methodology (TCM) as a valuation technique (Ariza *et al.*, 2012). TCM computes individuals' willingness to pay for participating in a given recreation activity (i.e. visiting a beach) by taking into account the costs (e.g. travel costs, access fees, equipment costs, or the opportunity cost of time) incurred by the individuals to participate in the activity. For the central beach of Lloret de Mar, the total consumer surplus in one day at the peak of summer, using the number of daily users estimated in Sardá *et al.* (2009), was computed as 1.23 million euros.

Doing an exercise of a low estimation (30 similar days during the summer peak and the same amount, 30 days during the rest of the year) we obtained an annual value of 73.8 million euros for this beach. Nineteen percent of this money (13.3 million euros) went into taxes received by different administrations.

The results also show an important gap between investments made by managers (less than 1 million euros for all municipal beaches during this year) and users' economic valuation (more than 1 million euros per day at the peak of the season). With all these data, the value per meter square of the central beach of Lloret de Mar was computed as 1320 euros and its annual value per ha on 13.2 million euros.

Methodologies used for estimating non-marketed resources present limitations and still we have a long way to go before general acceptance of these values. These methodologies need to be calibrated using information generated with other techniques. In the case of TCM, values are highly site dependent.

A similar exercise done in beaches of Southern New England (Kline and Swallow, 1998) yield a value an order of magnitude lower than the one computed for Lloret de Mar (93,536 US$ per ha). Brenner *et al.* (2010) computed the recreational value of beach ecosystem services using the Value Transfer approach; in this case, the value for the central beach of Lloret de Mar was calculated as 36,687 US$ per ha. A contingent valuation exercise done in the beach of S'Abanell (Blanes, Girona) just 5 kilometers south of the central beach of Lloret de Mar yield an annual value of 315,864 euros, (13,375 beach users per day; 16.4% willing to pay 2.4 euros per day as average; 30 similar days during the season and another similar 30 days computed during the rest of the year) which accounts for a value of 63,811 euros per ha. (Lozoya, 2012).

Obviously all these estimates need to be handled with care because they are clearly methodologically dependent but they just present information about the value of the beach as a provider of ecosystem services for people.

Ecosystem Services: A Business Policy Oriented Approach

Following the publication of the Millennium Ecosystem Assessment report (MEA, 2005), a new social-ecological paradigm have emerged to change the way we relate with natural systems. These systems are starting to be seen much more as capital assets that provide a flow of natural goods and ecosystem services to man. This view deals with the identification and analysis of the ecosystem services provided by nature but also with its valuation. Here is when the ecosystem service concept, if correctly explained, becomes highly useful to convince people about the importance to protect such sources of man's welfare, the ecosystems that are providing with these services to us. The ecosystem service concept also offer the possibility for changing the way we protect nature, moving from traditional conservational positions, many times seen by people as contra posed to development, towards a logic of conservation for development (Folke, 2006; Gómez-Baggethun and Ruiz-Pérez, 2011). Although this view is receiving more attention in the northern part of Europe where interdisciplinary collaboration to look things differently it is more rooted, Mediterranean countries should not avoid this particular way to see the future, a way that can be seen in a couple of examples.

Seagrass beds

Seagrass beds provide a high number of ecosystem services from provisioning, supporting, regulating and cultural services (Barbier *et al.*, 2011), they protect the coastal area from erosion, they serve as source of carbon sequestration, they are able to purify waters by concentrating pollutants without any apparent adverse effects, they constitute nursery ground for many species of fishes, and more. In many forms people is benefiting from such type of services. However, in the Mediterranean we are losing seagrass beds in an alarming way. In 1986 we estimated a total area for *Posidonia oceanica* beds of 6.8 million ha (Bethoux and Copin Montegut, 1986). Fifteen years later,

Buia *et al.* (2000) estimated this area between 2.5-5.5 million ha; a loss of 1.3-4.3 million ha in a short time. We do not have proper economic figures of such loss but an economic service valuation done recently indicates that the value lost is going to be very high (Emma Jackson personal communication) and the possibilities for recovering this value is minimal due to the low intrinsic growth of the *Posidonia oceanica* species.

Beaches

Beaches play a key role in the Mediterranean environment. However beaches are being eroded today in a very fast manner (Eurosion, 2004). Urbanization of the coast has turned coastal erosion from a natural phenomenon into a problem of growing intensity. In Catalonia, a recent study (CIIRC, 2010) showed that Catalonian beaches (excluding the ones of the Ebro Delta) are eroded at a rate of 1.07 m per year. If we take into consideration half of the value calculated for the central beach of Lloret de Mar (660 euros per meter square) as an average value for the Catalan coast, Catalonia would be losing every year 136.5 millions of euros as a consequence of erosion problems by resource disappearance. Obviously, tourist expenditure of not using this space can be substituted by another activity or by occupying other parts of the beach when beaches are still width enough, but if this trend is not reverse, it will reach one point in which such losses could just be considered irreversible and its values lost forever.

Most people believe that public institutions (sometimes non-governmental organizations) should take care about public goods and should be leading conservation planning for ecosystem services. As ecosystem services could go hand by hand with biodiversity protection (Chan *et al.,* 2006), public institutions could just take the potential to manage both things together. In this case, the impact that man (as individuals or as companies) made on the environment should be regulated by those institutions to take proper care of its use and abuse. Although it sounds logical, this is not working properly due, in most cases, to a lack of priorities in the protection of public goods by the authorities, and if we do not change the course of our actions, we may end up without anything to take care. The ecosystem services concept can be also useful at that point because it emphasizes the real notion of both protecting nature and benefiting man at the same time. We need to modify our way of thinking in our relation with natural systems and its conservation. Nature, in this case the Mediterranean Sea, brings us benefits through ecosystem goods and services, and, in a way, whoever benefit from those goods and services should have a responsibility of its degradation. Solutions for these problems require much more public-private partnerships and the involvement of all actors to come up with fresh ideas to protect those natural resources in a sustainable way.

Innovative schemes for linking public and private efforts to protect ecosystems by ensuring the provision of ecosystem services are becoming more and more used today. Some research suggests that some schemes like the "payment for ecosystem services (PES)" could be a fundamental move in this way (Wendland *et al.,* 2009; Farley and Costanza, 2010; Farley *et al.,* 2010). PES schemes are voluntary transactions where a well-defined ecosystem service is being "bought" by at least one buyer from (a minimum of one) ecosystem service provider if, and only if, the ecosystem service provider secures its provision. Although we can find the utilization of some of these schemes worldwide (almost all of them are done for terrestrial environments) their real impact on ecosystem services is still negligible. PES schemes differ greatly from conventional markets to address negative externalities of our activities following the "polluter pays principle" by the introduction of the "steward earns principle" in which positive externalities are addressed from the very beginning (Gómez-Baggethun and Ruiz-Pérez, 2011).

The use of payment schemes into the marine domain remains in its infancy with small movements today. One good example could be the "blue carbon" idea. Wetland areas and seagrass beds are common in the Mediterranean. They provide plenty of ecosystem services to people from which we benefit. These natural systems have demonstrated a large capacity for carbon sequestration in both the dominant plants and the sediments immediately below (Mateo *et al.*, 1997; Kennedy *et al.*, 2010). The Blue Carbon International Scientific Working Group, under UNESCO Patronage (Pidgeon *et al.,* 2011) have built a program to coordinate and guide establishment of coastal "blue carbon" as a conservation and management tool contributing to climate change mitigation and the development of associated conservation financing mechanisms. Assessing its feasibility, providing implementable recommendations, diminishing data gaps, designing and implementing programs of action, analyzing its possible use in voluntary schemes of carbon markets,… these are actions that require a bunch of work for as many actors as possible including private organizations. The protection of wetland areas and seagrasses is vital for society due to the provision of those different ecosystem services. The carbon sequestration service provided by those ecosystems can be marketed through a PES scheme. Several projects are right now trying to put forward the idea of introducing such type of frameworks for these ecosystems into voluntary or mandatory carbon markets through Clean Development Mechanisms (CDM) where different stakeholders would have the possibility to work on conservational strategies while they are benefiting themselves in a marketed way.

More conservation efforts in the fisheries industry need to be taken as soon as possible because in the Mediterranean, as well as in other parts of the world, fisheries are collapsing. The ability to maintain fishermen's income levels by protecting fish populations should be an essential element for success. In this case, a PES scheme also could work by providing an economic incentive to fishermen to adopt fisheries and/or management practices favorable to the sustainability of fish stocks. In the problem of fish depletion, doing nothing (no taking actions) or moving business to other sites are no solutions for the problem. Working with regulatory tools alone have been demonstrated many times to fail due to a lack of enforcement, so providing incentives by using public-private partnerships could be feasible and it have been proven possible as in the Chilean artisanal fisheries (Castilla and Defeo, 2005; Gelcich *et al.*, 2008). This requires understanding fishermen, understanding the science behind the problem, matching both by using a participatory process, and finally introducing an incentive package that give security to fishermen to cover its living styles while the restoration of fish stocks takes place using adequate co-management tools.

Working with ecosystem services also incorporate the need to properly account for the valuation and management schemes of single ecosystem services or of multiple ecosystem services. For example, clean drinking water, sediment transport, and coastal productivity are all benefits arising from a well-managed river ecosystem, but normally these services are managed independently in isolation one from another and tradeoffs over management options depend on partial negotiations and lobbying. In addition, sediment transport is an intermediate service for the final service which is having a safe recreational space in a beach to lay-down in summer time which is basic for the maintenance of the tourism industry. As we do not properly account for these relationships, many times we end up with decisions that are not the best ones for the entire society. Beneficiaries of the latest service can be local actors such as hoteliers or global actors such as tour-operators, but, in any case, they are normally far from the room in which important decisions that affect their business are taken on the provision of ecosystem services that benefit them.

It is time to develop proper structured methodologies that help people to understand, valuate, and manage risks and opportunities arising from our dependence to ecosystems and the provision of its services. It is not just public managers and non-profit organizations that should do the job, private companies and private managers should be aware of the status of such ecosystem services on which they depend for its future taking part on its protection. Recently, United Nations, an organization that usually only was talking with national bodies, have made large efforts to start to talk directly with other stakeholders such as companies in the protection of natural systems (MEA, 2005). Different organizations are incorporating this message into its portfolio of actions (TEEB, 2010; WBCSD, 2011; WBCSD-MI-WRI, 2012).

The World Business Council for Sustainable Development (WBCSD) together with the Meridian institute and the World Resource Institute developed recently a Corporate Ecosystem Service guidelines for identifying business risks and opportunities arising from ecosystem change. This methodology is aimed to help companies understand their dependence and impact on ecosystems and the resulting business risks and opportunities in a coherent, systematic manner (WBCSD-MI-WRI, 2012). By choosing the boundaries for the study, making an identification and analysis of the ecosystem services used by them, and developing strategies to address risk and opportunities connected with the trends observed for those ecosystem services, companies can benefit as well man does.

Public-private partnership alliances need to be developed to find the way to manage correctly ecosystem services. Contrarily to what it can be seen in the terrestrial environment, given the difficulties in sampling and understanding the processes occurring in the marine environment, comparatively little is known of how marine biodiversity contributes to the delivery of these goods and services relied upon by people (Austen et al., 2008). However, we can take examples from such type of actions done on carbon and water related services of watersheds (www.ecosystem marketplace.com) . Although the participation of the private sector in the protection and payment for ecosystem services in the marine domain is in its infancy, the potential for improvement and growth is large. We still need to work hard to inform better all parts of society; however the business community is recognizing the importance of ecosystem health and ecosystem functions to provide goods and services that are necessary to maintain the competitiveness and sustainability of business operations, being more and more involved with time in the effort to guarantee and adequate functioning of natural systems.

CONCLUSION

The relationship of Mediterranean people with the sea still remains close after 4000 long years of history. In its northern rim, coastal and marine environments have become the most popular tourism destination for the entire world. In the southern rim people is changing the remarkable resilient stability reached during last decades by emphasizing needs for democracy and development asking for more plentiful reforms. However, as more people get into coastal areas and more activities are observed there, more habitat is lost, more seafood need to be extracted to satisfy local demand, more highways in the sea are open to transport goods and people, and our appetite for its colonization to extract sand, minerals and other materials or to satisfy our demand for energy goes up. All these uses (food, transport, dumping rubbish, recreation, ...) have compromised the ability of the sea to operate correctly and put in danger some of its observed process, its structural units, and their functions and with them, the services that benefit man. The sea's functioning has been stretched beyond its limit and today everybody is starting to know what are the ecological, social and economic implications of this degradation. Understand the economic benefits of a healthy environment could be also useful to empower marine European policy prioritizing proactive actions to avoid further environmental degradation. Better information to allow people to understand about ecosystem services and to have well-formed preferences is one of the first goals to be reached in a near future in this area of knowledge.

As other worldwide coastal and marine environments, the ecosystems of the Mediterranean Sea provide ecological functions that directly or indirectly translate into benefits to humans through resources (natural goods and ecosystem services) that can be renewable and sustainable if properly managed. However those resources often are taken for granted and commonly are overused; even for important and dependent economic activities such as fisheries and tourism, efficient management and sustainable exploitation can be considered as exceptions rather than normal rules leading to resources depletion and collapse. In order to protect these resources, a much more functional-based approaches to ecosystems need to be developed to help identify the relevant dependences of ecosystem services provision. These approaches should provide qualitative and quantitative analytical capabilities of elements, processes and services which are responsible for human well-being as a way to translate the ecological complexity into a structure useful in natural resource management. Understanding of provision of ecosystem services (quantification), understanding of the benefits to human well-being from ecosystem services (valuation), and creating incentives for the sustainable provision of such ecosystem services (policies, good governance, alliances,...) should be recognized as a precondition for a sustainable future of our seas and coasts. The development of information-based platforms to understand these relationships can be observed as a second important goal to protect ecosystem services.

The approaches above have been recognized in the new European marine policy. In this policy, Good Environmental Status (GEnS), becomes the final desired vision to be reached using an ecosystem approach to management. The preamble 8 of the Marine Strategy Framework Directive (MSFD) mentions "*by applying and ecosystem-based approach to the management of human activities while enabling a sustainable use of marine goods and services, priority should be given to achieving or maintaining good environmental status in the Community's marine environment to continuing its protection and preservation, and to prevent subsequent deterioration*". Following this policy, new tools are developed to avoid global degradation of ecosystems and the services they provide. New managerial standards as the Ecosystem-Based Management System-EBMS (Sardá *et al.*, 2010) or the Corporate Ecosystem Service guidelines (WBCSD-MI-WRI, 2012), between others, are just some of the tools we are developing to help us in the move to protect ecosystems through the protection of the services they are given to us.

Valuing Nature by the natural goods and ecosystem services that provide is not new but incorporating monetary terms into this recognition it is. Introducing market-based instruments for conservation (Paterson *et al.*, 2010) have still detractors and it is still in debate; however, this issue has opened the language used in biological conservation to other stakeholders facilitating a common jargon to advance into sustainable ecosystem management models. Valuation of non-marketed ecosystem services need to be made with care because different techniques can produce different monetary outputs, and we are just now at the beginning of this new way to see nature, the possibility to give unit values to something was not valued before, opened the conservation movement to other actors making this much more relevant for people. Accounting for ecosystem services is important also for public policy because these services contribute significantly to human welfare and are not captured in existing welfare accounts (Boyd and Banzhaf, 2007). This opens a large research need to come up with a clear structure of what can be economically quantified as ecosystem service, in which way sand availability in a beach give us an economic value independently of all indirect multiplier posterior effects this service has in economy?. The development and application of valuation techniques that can be accepted by society in public accounts is also a fundamental objective to reach in a near future.

The importance of coastal and marine ecosystems for future development is enormous and the ecosystem service concept can be useful to move us in this recognition. In order to properly value ecosystems, a large interdisciplinary collaboration is required; ecologists, economists, sociologists, and others, need to work together in a high intellectual way to develop procedures that can facilitate the introduction of this protective schemes in daily societal operations. Still we have too much terminology with too little understanding of it, and few real proactive actions. If, in addition, we need to develop private-public collaborations, this used terminology and language styles (scholar, business, society,…) need to come up with something standard and simple to allow the analysis of integrated decision-making processes and solutions. Protecting the marine environment like the one in the Mediterranean (from small organisms to large ecosystems) is necessary, not only for its own health but also for our people's welfare. The new framework opened by the introduction of the ecosystem service concept into the management of public goods should help us to advance into the conservation of natural systems.

Acknowledgments

This work was carried out within the framework of the KNOWSEAS-FP7 project of the European Union and the MeVaPlaya-II project (CSO2009-14589-C03) of the National Research Plan of Spain.

References

Ariza, E., Ballester, R., Rigall, R., Saló, A., Roca, E., Villares, M., Jimenez, J.A. and Sardá, R. (2012). On the relationship between quality, users perception and economic valuation in the NW Mediterranean beaches. *Ocean and Coastal Management*, 63: 55-66.

Austen, M.C., Burrows, M., Frid, C., Haines-Young, R., Hiscock, H., Moran, D., Myers, J., Paterson, D.M. and Rose, P. (2008). Marine biodiversity and the provision of goods and services: identifying the research priorities. Report to the UK Biodiversity Research Advisory Group.

Barbier, E.B., Hacker, S.D., Kennedy, C., Koch, E.W., Stier, A.C. and Silliman, B.R. (2011). The value of estuarine and coastal ecosystem services. *Ecological Monographs*, 81: 169-193.

Bateman, I.J., Jones, A.P., Lovett, A.A., Lake, I.R. and Day, B.H. (2002). Applying geographical information systems (GIS) to environmental and resource economics. *Environmental Resource Economics*, 22:219–269.

Berkes, F. and Folke, C. (1998). Linking social and ecological systems for resilience and sustainability. In: F. Berkes and C. Folke (Eds.), Linking social and ecological systems: management practices and social mechanisms for building resilience. pp:15-25. Cambridge University Press. Cambridge.

Bethoux, J.O. and Copin-Montegut, G. (1986). Biological fixation of atmospheric nitrogen in the Mediterranean Sea. *Limnology and Oceanography*, 31: 1353-1358.

Blakemore, F. and Williams, A. (2008). British Tourists' Valuation of a Turkish Beach Using Contingent Valuation and Travel Cost Methods. *Journal of Coastal Research*, 246, 1469-1480.

Borja, A., Elliott, M., Carstensen, J., Heiskanen, A.S. and van de Bund, W. (2010). Marine management: Towards an integrated implementation of the European Marine Strategy Framework and the Water Framework Directives. *Marine Pollution Bulletin,* 60: 2175-2186.

Boyd, J. and Banzhaf, S. (2007). "What are ecosystem services?". *Ecological Economics*, 63: 616-626.

Brenner, J. (2007). Valuation of ecosystem services in the Catalan coastal zone. Doctoral Thesis Politécnica University of Barcelona. Barcelona, 186 pp.

Brenner, J., Jimenez, J.A. and Sardá, R. (2006). Definition of homogeneous environmental management units for the Catalan coastal zone. *Environmental Management*; 38: 993–1005.

Brenner, J., Jimenez, J.A., Sardá, R. and Garola, A. (2010). An assessment of the non-market value of the ecosystem services provided by the Catalan coastal zone, Spain. *Ocean and Coastal Management*, 53: 27-38.

Buia, M.C., Gambi, M.C. and Zupo, V. (2000). Structure and functioning of Mediterranean Seagrass ecosystems: an overview. *Biologia Marina Mediterranea*, 7: 167-190.

Castilla, J.C. and Defeo, O. (2005). Paradigm shifts needed for world fisheries. *Science,* 309: 1324–1325.

Chan, K.M.A., Shaw, M.R., Cameron, D.R., Underwood, E.C. and Daily, G.C. (2006). Conservation Planning for Ecosystem services. PLOS Biology, 4(11): e379.

Chapin, F.S.; Pamela A.M.; and Mooney, H.A.. (2002). Principles of Terrestrial Ecosystem Ecology. New York: Springer.

Centre Internacional d'Investigació dels Recursos Costaners (CIIRC). (2010). Estat de la Zona Costanera a Catalunya. Departament Política Territorial i Obres Públiques. Barcelona.

Coll, M., Piroddi, C., Steenbeek, J., Kaschner, K., Ben Rais Lasram, F., Aguzzi, J., Ballesteros, E., Bianchi, C.N., Corbera, J., Dailianis, T., Danovaro, R., Estrada, M., Froglia, C., Galil, B.S., Gasol, J.M., Gertwagen, R., Gil, J., Guilhaumon, F., Kesner-Reyes, K., Kitsos, M.S., Koukouras, A., Lampadariou, N., Laxamana, E., López-Fé de la Cuadra, C.M., Lotze, H.K., Martin, D., Mouillot, D., Oro, D., Raicevich, S., Rius-Barile, J., Saiz-Salinas, J.I., San Vicente, C., Somot, S., Templado, J., Turon, X., Vafidis, D,, Villanueva, R. and Voultsiadou, E. (2010). The Biodiversity of the Mediterranean Sea: Estimates, Patterns, and Threats. PLoS ONE 5:e11842.

Coll, M., Piroddi, C., Albouy, C., Ben Rais Lasram, F., Cheung, W.W.L., Christensen, V., Karpouzi, V.S., Guilhaumon, F., Mouillot, D., Paleczny, M., Palomares, M.L., Steenbeek, J., Trujillo, P., Watson, R. and Pauly, D. (2012). The Mediterranean Sea under siege: spatial overlap between marine biodiversity, cumulative threats and marine reserves. *Global Ecology and Biogeography*, 21:455-480.

Cooper, P. (2012). The DPSWR Social-Ecological accounting framework: notes on its definition and application. Policy Brief nº 3 EU FP7 KNOWSEAS project. ISBN 0-9529089-5-6.

Costanza, R., d'Arge, R., de Groot, R., Faber, S., Grass, M., Hannon, B., Limburg, K., Naeem, S., O'Neil, R.V., Paruelo, J., Raskin, R.G., Sutton, P. andvan den Belt, M. (1997). The value of the world's ecosystem services and natural capital. *Nature*, 387:253–260.

Costanza, R. and C. Folke (1997). Valuing ecosystem services with efficiency, fairness, and sustainability as goals. In: Daily, G.C. (Eds.) *Nature's services*. pp: 49-68. Island Press. Washington.

Crutzen, P.J. (2002). Geology of mankind: the anthropocene. *Nature*, 415: 23.

Daily, G.C. (1997). Nature's Services. Island Press, Washington. 393 pp.

Daly, H.E. (1992). Allocation, distribution, and scale: towards an economics that is efficient, just, and sustainable. *Ecological Economics*, 6: 185-193.

Departament d'Agricultura, Ramaderia i Pesca (DARP). (2000). Bathymetry of the Catalan coast up to 1000 m, 1:50,000 scale. Barcelona: Departament d'Agricultura, Ramaderia i Pesca, Generalitat de Catalunya. http://www.gencat.net/darp/c/pescamar/sigpesca/csig09.htm.

Departament d'Agricultura, Ramaderia i Pesca (DARP). (2002). Cartography of seagrasses. Barcelona: Departament d'Agricultura, Ramaderia i Pesca, Generalitat de Catalunya; 2002. http://www.gencat.net/darp/c/pescamar/sigpesca/csig07.htm.

De Groot, R. (1992). Functions of nature: evaluation of nature in environmental planning, management and decision making. Netherlands: Wolters-Noordhoff. 315 pp.

De Groot, R. (2006). Function-analysis and valuation as a tool to assess land-use conflicts in planning for sustainbable, multi-functional landscapes. *Landscape and Urban Planning,* 75: 175-186.

De Groot, R., Wilson, M.A. and Boumans, R.M.J. (2002). A typology for the classification, description and valuation of ecosystem functions, goods and services. *Ecological Economics,* 41: 393–408.

Departament de Medi Ambient i Habitatge (DMAH). (2006). Cartography of the habitats of Catalonia (CHC50), 1:50,000 scale. Barcelona: Departament de Medi Ambient i Habitatge, Generalitat de Catalunya. http://mediambient.gencat.net/cat/el_medi/habitats/habitats_cartografia.htm#cd.

Ehrlich, P. and Ehrlich, A. (1981). Extinction: the causes and consequences of the dissapearence of species. New York. Random House.

Ehrlich, P., Ehrlich, A. and Holdren, J. (1977). Ecoscience: Population, resources, Environment. San Francisco. W.H. Freeman.

European Environmental Agency (EEA). (1999). Environmental indicators: Typology and overview. Technical report nº 25. EEA, Copenhagen.

Eurosion (2004). Living with coastal erosion in Europe: sediment and space for sustainability. Eurosion consortium (www.eurosion.org). 54 pp.

Ewing, B., Moore, D., Goldfinger, S., Oursler, A., Reed, A. and Wackernagel, M. (2010). The Ecological Footprint Atlas 2010. Oakland, California. Global Footprint Network.

Farber, S., Costanza, R., Childers, D.I., Erickson, J., Gross, K., Grove, M., Hopkinson, C.S., Kahn, J., Pincentl, S., Troy, A., Warren, P. and Wilson, M. (2006). Linking ecology and economics for ecosystem management. *BioScience*, 56: 117-129.

Farley, J. and Costanza, R. (2010). Payments for ecosystem services: from local to global. *Ecological Economics*, 69: 2060-2068.

Farley, J., Aquino, A. Daniels, A., Moulaert, A., Lee, D. and Krause, A. (2010). Global mechanisms for sustaining and enhancing PES schemes. *Ecological Economics*, 69: 2075-2084.

Farmer, A., Mee. L., Langmead, O., Cooper, P., Kannen, A., Kershaw, P. and Cherrier, V. (2012). The Ecosystem Approach in Marine Management. EU FP7 KNOWSEAS Project. ISBN 0-9529089-5-.

Fisher, B. and Turner K. (2008). Ecosystem services: classification for valuation. *Biological Conservation*, 141: 1167-1169.

Folke, C. (2006). The economic perspective: Conservation against development versus conservation for development. *Conservation Biology*, 20: 686-688.

Gilbert A. J. and Janssen, R. (1998). Use of environmental functions to communicate the values of a mangrove ecosystem under different management regimes. *Ecological Economics*, 25: 323-346.

Gelcich, S., Godoy, N., Prado, L. and Castilla, J.C.. (2008). Add-on conservation benefits of marine territorial user rights fishery policies in Central Chile. *Ecological Applications*, 18: 273-281.

Gómez-Baggethun, E., de Groot, R., Lomas, P.L. and Montes, C. (2010). The history of ecosystem services in economic theory and practice: From early notions to markets and payment schemes. *Ecological Economics*, 69: 1209-1218.

Gómez-Baggethun, E. and Ruiz-Pérez, M. (2011). Economic valuation and the commodification of ecosystem services. *Progress in Physical Geography*, 35: 613-628.

Gren, I.M. and Soderqvist, T. (1994). "Economic valuation of wetlands: a survey". Beijer International Institute of Ecological Economics. The Royal Swedish Academy of Sciences, Stockholm

Gurluk, S. and Rehber, E. (2006). A travel cost study to estimate recreational value for a bird refuge at Lake Manyas, Turkey. *Journal of Environmental Management,* 88: 1350-*1360.*

Halpern, B.S., Walbridge, S., Selkoe, K.A., Kappel, C.V., Micheli, F., D´Agrosa, C., Bruno, J.F., Casey, K.S., Ebert, C., Fox, H.E., Frujita, R., Heinemann, D., Lenihan, H.S., Madin, E.M.P., Perry, M.T., Selig, E.R., Spalding, M., Steneck, R. and Watson, R. (2008). A Global Map of Human Impact on Marine Ecosystems. *Science*, 319: 948-952.

Kennedy, H., Beggins, J., Duarte, C., Fourquean, J.W., Holmer, M., Marbá, N., and Middleburg, J.J.. (2010). Seagrass sediments as a global carbon sink: isotropic constraints. *Global Biogeochemical. Cycles.*, 24: 1-8.

Kline, J.D. and Swallow, S.K. (1998). The demand for local access to coastal recreation in Southern New England. *Coastal Management*, 26: 177-190.

lberini, A. and Rosato, P. (2004). Information and willingness to pay in a contingent valuation study: the value of S. Erasmo in the Lagoon of Venice. Fondazione Eni Enrico Mattei (FEEM) Note di Lavoro Working Paper nº 19-2004.

Lozoya J.P. (2012). Multi-risk assessment and user's perception: a further step towards ecosystem-based beach management. Doctoral Dissretation. Universidad Politécnica de Catalunya. 187 pp.

Mateo, M., Romero, J., Pérez, M., Littler, M.M. and Littler, D.S. (1997). Dynamics of millenary organic deposits resulting from the growth of the Mediterranean seagrass *Posidonia oceanica*. *Estuarine Coastal and Shelf Science*, 44: 103-110.

Millennium Ecosystem Assessment (MEA). (2005). Ecosystems and Human Well-being Synthesis. Washington DC. Island Press.

Nunes, P. and Rossetto, L. (2004). Measuring the economic value of alternative clam fishing management practices in the Venice Lagoon: results from a conjoint valuation application. *Journal of Marine Systems,* 51:309-320.

Nunes, P. and Silvestri, S. (2008). Regulation of the fishing activities in the lagoon of Venice, Italy: Results from a socio-economic study. *Estuarine, Coastal and Shelf Science,* 80: 173-180.

Peters, H. and Hawkins, J. (2009). Access to marine parks: A comparative study in willingness to pay. *Ocean and Coastal Management*, 52: 219-228.

Pidgeon, E., Crooks, S., Tamelander, J. and Fonseca, L. (2011), Blue Carbon" International Scientific Working Group. (http://www.marineclimate change.com/marineclimatechange/bluecarbon_2.html).

Ragkos, A. and Psychoudakis, A. (2006). Using a functional approach to wetland valuation: The case of Zazari-Cheimaditida. *Regional Environmental Change*, 6: 193-200.

Rockström, J., Steffen, W., Noone, K., Persson, A. Stuart Chapin, F., Lambin, E.F., Lenton'T.M., Scheffer, M., Folke, C., Schellnhuber, H.J., Nykvist , B., de Wit, C.A., Hughes, T., van der Leeuw, S., Rodhe, H., Sörlin, S., Snyder, P.K., Costanza, R., Svedin, U., Falkenmark, M., Karlberg, L., Corell, R.W., Fabry, V.J., Hansen, J., Walker, B. Liverman, D., Richardson, K., Crutzen, P. and Foley, J.A.. (2009). A safe operating space for humanity. *Nature*, 461: 472-475.

Sala, E., Ballesteros, E., Dendrinos, P., Di Franco, A., Ferretti, F., Foley, D., Fraschetti, S., Friedlander, A., Garrabou, J., Güçlüsoy, H., Guidetti, P., Halpern, B.S., Hereu, B., Karamanlidis, A.A., Kizilkaya, Z., Macpherson, E., Mangialajo, L., Mariani, S., Micheli, F., Pais, A., Riser, K., Rosenberg, A.A., Sales, M., Selkoe, K.A., Starr, R., Tomas, F. and Zabala, M.(2012). The Structure of Mediterranean Rocky Reef Ecosystems across Environmental and Human Gradients, and Conservation Implications. *PLoS ONE,* 7: e32742.

Sardá, R. (2001). Shoreline development on the Spanish Coast. pp: 149-165. In: von Bodungen, B and Turner, R.K. (Eds). Science and Integrated Coastal Management. Berlin. Dahlem University Press.

Sardá, R. and Fluvià, M. (1999). Tourist development in the Costa Brava (Girona, Spain): a quantification of pressures on the Coastal Environment. pp: 257-277. In: Salomons, W., Turner, R.K., Lacerda, L.D. and Ramachandran, S. (Eds). Perspectives on Integrated Coastal Management. Berlin. Springer Publ.

Sardá, R., Mora, J., Ariza, E., Avila, C. and Jimenez, J.A.. (2009). Decadal shifts in beach user sand availability on the Costa Brava (Northwestern Mediterranean coast). *Tourism Management*. 30: 158-168.

Sardá, R. and col. (2002). L'Espai Natural de Pinya de Rosa: informe científic. Generalitat de Catalunya. 179 pp.

Stithous, M. (2009). Respondent Certainty and Payment Vehicle Effect in Contingent Valuation: an Empirical Study for the Conservation of Two Endangered Species in Zakynthos Island, Greece. University of Stirling Economics Discussion Paper 2009-21.

Steffen, W., Crutzen, P. and McNeill, J.R. (2007). The Anthropocene: are humans now overwhelming the great forces of Nature. *Ambio*, 36: 614-621.

Study on Critical Environmental Problems (SCEP). (1970). Man's Impact on the Global Environment. Cambridge, Mass. MIT Press.

The Economics of Ecosystem and Biodiversity (TEEB). (2010). The economics of of ecosystem and biodiversity: ecological and economic foundations. UNEP TEEB Office.

Togridou, A., Hovardas, T. and Pantis, J.D. (2006). Determinants of visitors' willingness to pay for the National Marine Park of Zakynthos, Greece. *Ecological Economics*, 60: 308-319.

Troy, A. and Wilson, M.A. (2006). Mapping ecosystem services: practical challenges and opportunities in linking GIS and value transfer. *Ecological Economics*; 60:435–449.

Wendland, K.J., Honzák, M., Portela, R, .Vitale, B., Rubinoff, S. and Randrianarisoa, J. (2009). Targeting and implementing payments for ecosystem services: opportunities for bunding biodiversity conservation with carbon and ater services in Madagascar. *Ecological Economics*, 69: 2093-2107.

Westman, W.E. (1977). How much are nature's services worth?. *Science*, 197: 960-964.

World Business Council for Sustainable Development (WBCSD). (2011). Guide to Corporate Ecosystem valuation. (http://www.wbcsd.org)

World Business Council for Sustainable Development (WBCSD), Meridian Institute (MI), World Resources Institute (WRI). (2012). The Corporate Ecosystem services Review. (http://www.wbcsd.org)

Worm, B., Barbier, E.B., Beaumont, N., Duffy, J.E., Folke, C., Halpern, B.S., Jackson, J.B.C., Lotze, H.K., Micheli, F., Palumbi, S.R., Sala, E., Selkoe, K.A., Stachowicz, J.J. and Watson, R. (2006). Impacts of Biodiversity Loss on Ocean Ecosystem Services. *Science*, 314: 787-790.

Yepes, V. (2004). La gestión de playas basándose en normas de calidad y medio ambiente. Proceedings of the II International Conference of Civil Engineering, Territory and Environment. Santiafgo de Compostela, Spain.

Zanatta, V., A. Alberini, P. Rosato, and Longo, A. (2005). The value of recreational sport fishing in the Lagoon of Venice: evidence from actual and hypothetical fishing trips. Paper presented at the EAERE Annual Meeting, Budapest.

In: Mediterranean Sea
Editor: Terrence B. Hughes

ISBN: 978-1-62618-238-7

Chapter 2

MARINE POLLUTION FROM SHIPWRECKS AT THE SEA BOTTOM: A CASE STUDY FROM THE MEDITERRANEAN BASIN

***Mario Sprovieri*[1*]*, Marco Barra*[2]*, Marianna Del Core*[1]*, Gabriella Di Martino*[2]*, Luigi Giaramita*[1]*, Serena Gherardi*[2]*, Sara Innangi*[2]*, Elvira Oliveri*[1]*, Salvatore Passaro*[2]*, Teresa Romeo*[3]*, Paola Rumolo*[2]*, Daniela Salvagio Manta*[1]*, Stella Tamburrino*[2]*, Renato Tonielli*[2]*, Anna Traina*[1]*, Giorgio Tranchida*[1]*, Mattia Vallefuoco*[2]*, Salvatore Mazzola*[1]* and Franco Andaloro*[3]**

[1]IAMC-CNR, Capo Granitola, Italy
[2]IAMC-CNR, Naples, Italy
[3]ISPRA, Milazzo, Italy

* Corresponding author: Dr. Mario Sprovieri. Institute for Coastal and Marine Environment (IAMC) – CNR. Via del Mare, 3, 91021 Torretta Granitola, Campobello di Mazara (TP), Italy. Tel: +39092440670; Fax: +39092440445; E-mail: mario.sprovieri@iamc.cnr.it.

ABSTRACT

In order to properly adopt the international recommendations on the marine environment and biodiversity conservation, as well as fishing resources and human health, an increasing effort is incoming in the last two decades to study, prevent and mitigate the anthropogenic environmental impact in the Mediterranean sea. The main research efforts and policy strategies implemented in the last two decades are regarding oil pollution (dumping, spills, shipping, extraction, accidents, etc.), civil and industrial waste discharges and, recently, littering. Less known are the other pollution sources generally not visible and frequently distributed offshore. One of this cryptic sources of pollution is represented by military and commercial shipwrecks abandoned on the sea bottom. Actually, fuels, construction materials, army and transported goods deposited at the bottom of the sea can generate ecological damage thus modifying natural biological diversity and fishing activity with unpredictable hazards and dangerous effects on the marine ecosystem. In particular, indiscriminate and uncontrolled leakage of toxic material and organic/inorganic pollutants from those vessels could affect the wellbeing of the deep sea also for long-time.

Very few data are available on the effects of pollution produced by shipwrecks in the Mediterranean sea although thousand of military ships were abandoned at the seabed of the basin as effect of accidents during the first and the second word wars.

This contribution aims to explore the potential effects of contamination of a number of shipwrecks identified by multi-proxy geophysical tools at the bottom of one of the key areas of the Mediterranean sea, around the Sicily island, where intense ship traffic and fishing activities occur. Geophysical exploration and geochemical data (heavy metals, PAHs, PCBs, Pesticides, hydrocarbons, etc.) are here presented for selected cases and testify the high risk represented by this kind of point-source pollution on the marine system.

1. INTRODUCTION

A sustainable and appropriate governance and management of the environment requires the availability of reliable and comprehensive information on the natural system as well as information on the social, economic, legal and political compartments. The ocean and in particular, the deep seafloor covers approximately 60% of Earth's surface and only a very limited part of it has been the focus of scientific investigation. Whilst remoteness and inaccessibility limited scientific and research actions, they have not protected these depths from human impacts. Increasing demand for living and non-living resources and diminishing or exhausted reserves on land and in shallow water are pushing human activities ever deeper into the world's oceans. At the same time advances in technology now allow access to resources of economic value that were previously inaccessible. This has resulted in an increasing number of direct and indirect anthropogenic pressures on deep-sea ecosystems. Governance and management of the deep sea is of increasing international concern. The United Nations, the Regional Seas conventions and organisations, including the European Union, are developing marine environment policies as well as monitoring and reporting procedures. Rules and codes of conduct are being established to regulate activities impacting on the deep ocean. For example, the OSPAR Commission has recognised the scientific case for establishing Marine Protected Areas in areas beyond national jurisdiction in the deep North East Atlantic (e.g., Ospar, 2008a). It has developed a code of conduct for Responsible Marine Research in the Deep Seas and High Seas of the OSPAR Maritime Area (Ospar, 2008b). Under a new memorandum of understanding, adopted by the two organisations in 2008, an attempt is being made to combine fisheries and conservation management (Ospar, 2008a). The requirement for environmental and socio-economic data is recognised in many political forums. The 1995 United Nations (UN) Fish Stocks Agreement calls for the sharing of "complete and accurate data concerning fishing activities" (UN, 1995). The Convention on Biological Diversity [UN, 1992] promotes the ecosystem approach as its primary framework for action. The ecosystem approach is a strategy for the integrated management of land, water and living resources that promotes conservation and sustainable use in an equitable way, recognizing that humans and their activities are integral to ecosystems. At the European level, the Marine Strategy Framework Directive (MSFD) (EC, 2008) and the OSPAR Biological Diversity and Ecosystems Strategy [16] both require assessments of human activities within the marine environment, some of which will be in the deep sea and beyond national jurisdictions. To fulfil these assessments and to implement the ecosystem approach, comprehensive and consistent information on human activities is necessary. Data on human activities are collected and held i) by public institutions and private companies to fulfill regulatory requirements, ii) for commercial and operational purposes and iii) or for scientific research.

The main research efforts and policy strategies carried out in the last two decades are regarding oil pollution (dumping, spills, shipping, extraction, accidents, shipping), civil and industrial wastes discharge and recently littering. Here we report on the potential role played by a number of shipwrecks identified at the bottom of the Mediterranean Sea and on the contamination of the surrounding bottom sediments in terms of organic and inorganic pollutants. Specifically, the studied area represents a spot zone where to properly explore multidisciplinary methodologies for assessment of anthropic impact at the bottom of the sea caused by different kinds of shipwrecks. There are thousand of vessels, aircrafts, and other commercial and military devices present at the bottom of the Mediterranean sea due to accidents, collisions, etc. Information by fishermen and dedicated surveys provided preliminary informative steps for a specific selection of a number of targets within the context of a national scientific program funded by the Sicilian Department of Agriculture and Fisheries and focused on the exploration of the potential effects of shipwrecks on the marine ecosystem and fisheries around the Sicilian coasts. Thus, in this research we try to explore the potential effects on marine pollution produced by leakage and contaminants dispersal from the studied shipwrecks on the surrounding sediments as first step of a larger and more definitive ecosystem impact. Specifically, organic and inorganic contaminants were investigated from samples collected around four shipwrecks identified and characterised by acoustic methodologies. Future researches, planned to investigate pathways of contaminants from the same point sources to the ecosystem and specifically to the fish compartments are on going and represent a further step of this intriguing story.

2. Materials and Methods

2.1. Sampling Area and Strategy for Sampling

Four shipwrecks were identified, characterised by acoustic remote surveys and sampled, by collection of sediments, on the platform of the western Sicily (Figure 1). In particular, the Target 1 (Kent) is located in the Castellammare gulf at a distance of 5.4 Km from the coast of the Castellammare city (NW Sicily), the Target 2 (Pavlov V) is located in front of the Trapani city (western Sicily) at a distance of about 3.6 Km from the coast, the Target 3 (Thetis) is located in front of the Mazara del Vallo city (southern Sicily) at a distance of about 8.5 Km from the coast and, finally, the Target 4 (unknown wreck) is located at ENE of the Pantelleria Island at a distance of about 38 Km from the coast and at a depth of 180 m. Sampling was performed during May 2010, on board of the R/V Urania.

Two sediment samples were collected in the surrounding area of the four shipwrecks as reported in Figure 1. Basically, a first sample is located within a distance of about 50 m from the wrecks and the other one is placed at a distance of about 100-200 m. The two samples should provide information on a first order sediment pollution produced by the vessel and with the second point we tried to document the natural background of the area. Obviously, we are aware that effects of contamination generated by the wrecks could be more widespread with respect to this basic sampling approach and therefore we also explored effects of contamination by direct comparison with larger scale background datasets.

Hereafter we report on a brief description of historical data available from the literature on the four shipwrecks.

The Kent cargo (Target 1) sunk near San Vito Cape in 1978 due to a fire started in the engine room. The shipload was made by 27 tons of copper, 1300 kg of oil, 20 tons of polyethylene (partially recovered), 1400 kg of cigarettes and 20 tons of fumigators. In addition, the Kent was also known as the "Koran Ship", as this was carrying hundreds of Koran books. The "Pavlov V" tanker (Target 2) was built in 1951 and sunk on January 1978 near to the Trapani harbour. The "Thetis" ship (Target 3) was an oceanographic vessel of the National Research Council (CNR) of Italy, that sunk four miles southward from the Mazara del Vallo (Trapani) coast in August 2007, as a consequence of the impact with a cargo. Finally, an unknown target (Target 4) is located at about 20 miles outward from the coast of the Pantelleria Island, at a depth of 180 m.

2.2. Acoustic Methodologies

The acustic survey of the studied area has been carried out by a Kongsberg EM710 (100 kHz) multibeam sonar system. The acquisition depth has been between 35 and 200 m. According to the International Hydrographic Organization standard (IHO, 2008), the acquisition was carried out along parallel track routes with at least 50% overlap. Sound speed profiles in the water column were acquired every 8 h. The acoustic footprint varied in size between about 0.5 up to 1.3 m.

2.2.1. Multibeam Echosounder

The bathymetric measurements have been performed with a hull-mounted Kongsberg EM710 multibeam echosounder (MBSE). The system operates at a frequency of 100 kHz and it provides a 150° swath coverage to a maximum depth of 2000 m (Table 1).

The Urania vessel was equipped with a 12 channel Landstar Differential Global Positioning System (GPS), a motion sensor and a Meridian Surveyor gyrocompass; a velocity sound probe was installed near the multibeam transducers thus providing the real-time sound speed for the beam-forming, while a SeaBird CTD was lowered through the water column every 6–8 hours to provide a sound velocity profile useful to the depth computation.

The Seafloor Information System software was used for recording bathymetric data; it receives data from the GPS system and from the motion sensor in order to perform a real-time correction of bathymetric data for shipwreck position and attitude (Figure 2); some parameters such as *Nadir angle, Pulse Lenght* and *Max Coverage*, can be modified during data acquisition in order to improve data quality. Moreover, Reson PDS2000 software was used as navigation software during sampling operations.

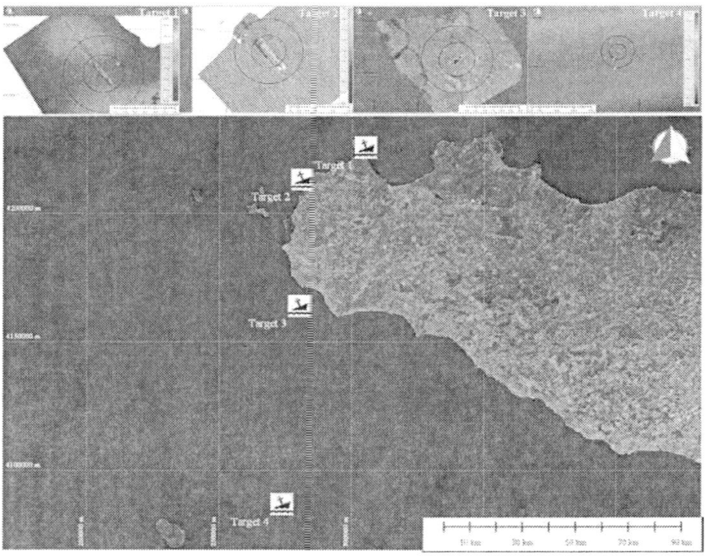

Figure 1. Location map of the studied shipwrecks and details of sampling sites (at the top) for each of them. Circles: distances of 50 and 200 m; White points: sampling sites around the vessels.

Backscatter data have been also provided from EM710, and the real-time *seabed image* display was useful for the target identification.

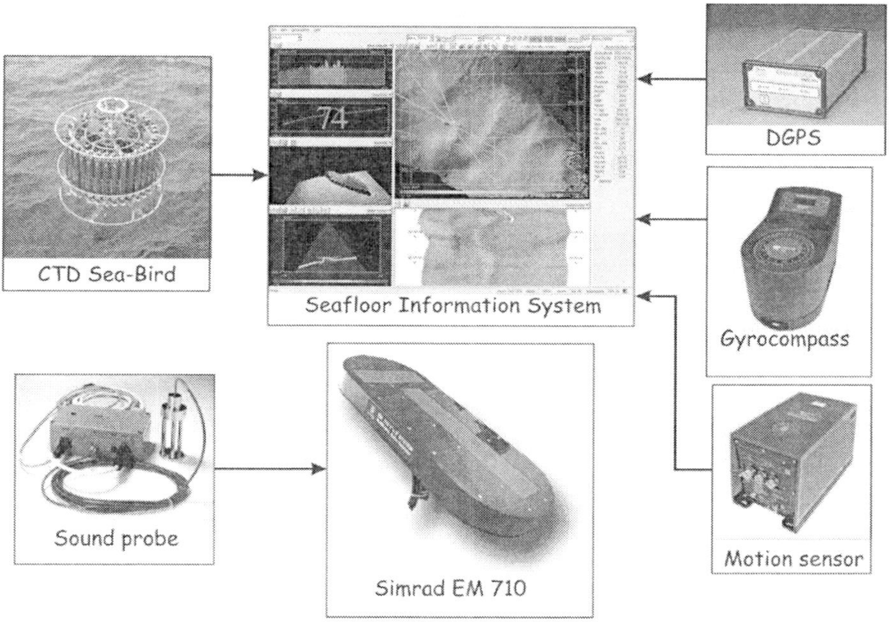

Figure 2. Seafloor Information System with devices used for geophysical data collection.

Table 1. Numerical setting of devices for geophysical data acquisition

EM 710 Multibeam	
Frequency range	70 to 100 kHz
Max ping rate	30 Hz
Swath coverage sector	Up to 140 degrees
Depth resolution	1 cm
Min depth	3 m below transducer
Max depth	2000 m
Fm sweep pulse	Max 200ms
Cw transmit pulses	0.15 to 2 ms
Roll stabilized beams	Yes, ±15°
Pitch stabilized beams	Yes, ±10°
Yaw stabilized beams	Yes, ±10°
Sounding patterns	Equiangular Equidistant HighDensity

2.2.2. Sidescan Sonar

The data from Sidescan sonar have been acquired after target identification trough multibeam data exploration. The EdgeTech 4200S operates at a double frequency of 100/500 kHz; it has been lowered in the water column up to 30 meter from the sea-bottom and an operating range of 150 m per channel was used during acquisition. Data from Sidescan sonar have been investigated for a limited number of targets because of the marine weather conditions and some logistic problems.

2.2.3. Processing of Geophysical Data

Bathymetric data have been processed by Caris Hips and Sips 7.0.2. software in order to produce high-resolution data. Tidal curves recorded in the proper time range have been applied to data, and then positioning data processed in order to remove uncorrected point caused by GPS signal loss or dilution. The main processing procedure consisted of a manual data cleaning using a 2D view and a 3D view of the swaths in order to remove bathymetric spikes. This procedure was utilized instead of a statistical data cleaning to preserve data resolution. The final bathymetric grids have been generated at variable resolution, 25 cm up to 5 meters grid cell, depending on the survey depth.

Data from Sidescan sonar have been processed by Triton Elics Isis and Delph Map software. A *Time Variable Gain* and a correct *bottom track* have been applied during the processing; navigation data have been also corrected performing a data smoothing and a *beam angle* correction, i.e. a statistical computation of the acoustic signal normalization curve. Finally sidescan data have been geo-referenced and a 50 cm resolution mosaic has been produced by Delph Map software.

Table 2. Geographical position and details about sediment samples near the four investigated shipwrecks

Shipwreck	Lat. N	Long. E	Samples	Sinking (year)
Target 1 (Kent)	38° 09' 31.8000"	12° 46' 37.7000"	2	1978
Target 2 (Pavlov V)	38° 02' 30.8000"	12° 30' 23.9000"	1	1978
Target 3 (Thetis)	37° 35' 40.4400"	12° 30' 43.9200"	2	2007
Target 4 (unknown)	36° 53' 52.4000"	12° 27' 43.0000"	2	-

2.3. Sampling for Geochemical Analysis

Sea bottom sediment samples were collected from the four areas around the identified shipwrecks (Figure 1), using a 20 l Van Veen grab. At each sampling sites two samples were collected, specifically at about 50 m and ~200 m from the shipwreck, expect for Target 3. Details of the sampling sites are reported in Table 2. Sediments were sub-sampled on-board, and tentatively refer to the first two centimetres of the bottom. These aliquots were homogenized with a plastic spoon, placed into pre-cleaned high-density polyethylene (HDPE) bottles and stored at -18°C on board, within an hour of collection, until the analysis.

Geochemical and mineralogical analyses were performed in the laboratories of the Institute for Coastal and Marine Environment (IAMC)-CNR of Capo Granitola and Naples.

2.3.1. Mineralogical Analyses

The bulk sample mineralogy was investigated on dried (T= 50°C) and powered samples by X-ray diffraction (XRD) using a D8 Discover Bruker, equipped with Sol-X detector, Cu-Kα radiation and scanning speed of 2 2°ϑ/min. The semi-quantitative analysis of minerals was performed according to methods and data reported in Schultz (1964) and Barahona et al. (1982).

2.3.2. Grain size analyses

Samples were treated with hydrogen peroxide (35%) and distilled water (1:4) for 48 hours at room temperature and then washed several times with distilled water and dispersed ultrasonically for 5 minutes before laser scattering analysis. Analyses were performed with a Horiba Partica LA-950V2 laser granulometer.

2.3.3. Total Organic Carbon (TOC)

Samples were weighted (15–20 mg) into Ag capsules and acidified with 1M HCl, in order to eliminate the carbonate fraction. To confirm the complete removal of inorganic carbon, samples were acified until effervescence was not detected in two consecutive acidification cycles (Nieuwenhuize et al., 1994; Schubert and Nielsen, 2000). Subsequently, the samples were dried in oven, at T= 60°C. Total organic carbon was measured on freeze-dried powdered samples by a Thermo Electron Flash Elemental Analyzer (EA 1112) (Verardo et al. 1990; Standard Operating Procedure 2007). External standard was acetanilide (C_8H_9ON, elemental composition: 71.09% carbon and 10.36% nitrogen). Samples were run with blank (empty silver capsule) and known acetanilide standards in order to correct the C associated with silver cups and calibrate for elemental analysis. Instrumental detection limits for TOC is 0.007%.

2.3.4. Major, Minor and Trace Elements

Aluminium, As, Ba, Cd, Co, Cr, Cu, Fe, Mn, Mo, Ni, V and Zn concentrations analysis were carried out by ICP-AES (Thermo iCAP 6000) after total acid digestion of the samples according to EPA 3052, 6010C methods. About 0.25 g of sediment, previously dried (at T= 40°C) and homogenized in a agata mortar, were placed in Teflon closed pressure vessels, with an acid mixture (6 ml HNO_3, 3 ml HCl and 2 ml HF), and heated at T=220±5 °C for 90 min in microwaves oven (Anton Paar Multiwave 3000). In order to minimize contamination risks, high purity grade reagents and acid-cleaned laboratory materials were used. Reference Standard Materials (PACS-2, BCR-2), reagent blanks and duplicated samples (about 20% of the total number of samples) were prepared and measured to assess accuracy (estimated between 86 and 105%), detection limits, precision (routinely better than 6%; RSD%, n=3) and reproducibility (better than 7% for all the elements) of the analyses. All results were calculated with respect to dry weight.

2.3.5. Polycyclic Aromatic Hydrocarbons (PAHs)

Only the PAHs indicated by the US-EPA Clear Water Act as priority ones, were analysed: PHE, phenanthrene; AN, anthracene; FLU, fluoranthene; PYR, pyrene; B(a)A, benzo[a]anthracene; CRY, chrysene; B(b)F, benzo[b] fluoranthene; B(k+j)F, benzo[k+j]fluoranthene; B(a)P, benzo[a]pyrene; B(e)P, benzo[e]pyrene; PER, perylene; IND, indeno[1,2,3-cd]pyrene; D(a,h)A, dibenzo[a,h]anthracene; B(g,h,i)P, benzo[ghi]perylene. Sediment samples were air dried and homogenised before the analysis. Two grams of sediment samples spiked with a 45 ng of surrogate standard of PAHs (SS; deuterated: phenantrene-d 10, fluoranthene-d 10, benzo[a]anthracene-d 12, benzo[a]pyrene-d 12, dibenzo[a,h]anthracene-d 14) were extracted with accelerate solvent extraction (DIONEX ASE 200) using hexane/acetone (80:20 v/v) mixture. The extraction cells were heated to 113°C until the pressure reached 1500 psi. The extracts, concentrated and re-dissolved with 1 ml of cyclohexane, were purified by SPE cartridges containing 6 g silica and eluted first with 10 ml of n-hexane and then with 20 ml cyclohexane:acetone (70:30) mixture. The final extracts were concentrated and re-dissolved with

400 μl solution of two deuterated PAHs (acenaphthylene-d 8 and chrysene-d 12) and finally analysed by capillary column gas chromatography with mass spectrometric detection. An analytical column DB-5 (95% dimethyl–5% diphenylpolysiloxane) 30 m x 0.25 mm x 0.25 mm was used. The temperature program was the following: 80°C (held for 1.5 min) to 200°C at 15°C/min, then 305°C (held for 10 min) at 7°C/min. The injection volume was 1 μl in splitless mode. Operative conditions were: ion source 250°C, inlet 250°C and carrier gas He 1.4 ml/min. The mass spectrometer was operated at an EI of 70 eV and in selective ion monitoring (SIM) mode. Laboratory quality control procedures included analyses of blanks, reference material and spiked samples. The reference material used for quality control was SRM 1941b-NIST. Instrument stability and response was checked using mix standard solutions. Estimated recoveries for each analyte ranged between 94% and 107%. Accuracy estimated on multiple analysis of the reference material was estimated better than ±10% for each single analyte. Reproducibility estimated on triplicate samples was >90%. The limit of detection was estimated as 1 μg/kg for each PAHs. All results were calculated with respect to dry weight.

2.3.6. Polychlorinated Biphenyls (PCBs)

Thirteen PCB congeners (28, 52, 77, 81, 101, 118, 126, 138, 153, 156, 169, 180, 128+157) were analysed. It implies the sum of PCBs (expressed as ΣPCB) is underestimated. Sediment samples were air dried and homogenised before the analysis, then 2 g together with a 1.0 ml solution of ^{13}C PCB-105 were extracted with accelerate solvent extraction (DIONEX ASE 200) using a hexane/acetone (80:20 v/v). The extraction cells were heated to 113°C until the pressure reached 1500 psi. The extracts concentrated and re-dissolved with a 0.5 ml hexane:isoctane (1:1 v/v) mixture, were then purified using Florisil (1 g) column and eluted with a 20 ml mixture hexane:isoctane (1:1 v/v). The final solutions were concentrated and re-dissolved with 400 μl internal standard solution of PCB-209 and finally analysed by capillary column gas chromatography with mass spectrometric detection. An analytical column DB-5 (95% dimethyl–5% diphenylpolysiloxane) 30 m x 0.25 mm x 0.25 μm was used. The temperature program was the fallowing: 80°C (held for 2 min) to 140°C at 18°C/min, then 300°C at 7°C/min. A volume of 1 μl was injected in splitless mode. Operative conditions were: ion source 280°C; inlet 250°C and carrier gas He 1.4 ml/min. The mass spectrometer was operated at an EI of 70 eV in MS-MS mode. Laboratory quality control procedures included analyses of sample blanks, reference material and spiked samples. The reference material used for quality control was SRM 1941b-NIST. Instrument stability and response was checked using NIST standard solutions. Estimated recoveries for each analyte ranged between 96% and 105%. Accuracy estimated on multiple analysis of the reference material was estimated better than ±10% for each single analyte Reproducibility estimated on triplicate samples was >95%. The limit of detection was estimated as 0.5 μg/kg for each PCBs. All results were calculated with respect to dry weight.

2.3.7. Organochloride Pesticide (OCPs)

Eighteen different organochloride pesticide congeners (α-hexachlorocycloexane, β-hexachlorocycloexane, γ-hexachlorocycloexane, heptachlor, heptachlor epoxide, aldrin, endrin, dieldrin, oxychlordane, cis-chlordane, trans-chlordane, cis-nonacloro, trans-nonacloro, DDE (o,p+p,p), DDD (o,p+p,p), DDT (o,p+p,p), mirex and methoxychlor) were analysed.

Sediment samples were air dried and homogenised before the analysis, then 2 g together with a 1.0 ml solution of α-hexachlorocycloesane-d6 were extracted with accelerate solvent extraction (DIONEX ASE 200) using a hexane/acetone (80:20 v/v). The extraction cells were heated to 113°C until the pressure reached 1500 psi. The extracts concentrated and redissolved with a 0.5 ml hexane:isoctane (1:1 v/v) mixture, were then purified using Florisil (1 g) column and eluted with a 20 ml mixture hexane:isoctane (1:1 v/v). The last solutions were concentrated and re-dissolved with 600 μl internal standard solution of DDE-d12 and finally analysed by capillary column gas chromatography with mass spectrometric detection. An analytical column DB-5 (95% dimethyl–5% diphenylpolysiloxane) 30 m x 0.25 mm x 0.25 μm was used. The temperature program was the fallowing: 80°C (held for 2 min) to 140°C at 18°C/min, then 300°C at 7°C/min. A volume of 1 μl was injected in splitless mode. Operative conditions were: ion source 250°C; inlet 280°C and carrier gas He 1.4 ml/min. The mass spectrometer was operated at an EI of 70 eV with technique of tandem mass spectrometry (MS-MS). Laboratory quality control procedures included analyses of sample blanks, reference material and spiked samples. The reference material used for quality control was SRM 1941b-NIST. Instrument stability and response was checked using mix standard solutions. Estimated recoveries for each analyte ranged between 96% and 105%. Accuracy estimated on multiple analysis of the reference material was estimated better than ±10% for each single analyte. Reproducibility estimated on triplicate samples was >95%. The limit of detection was estimated as 1 μg/kg for each pesticide. All results were calculated with respect to dry weight.

2.3.8. Diesel Range Organics (DROs)

Sediment samples were air dried and homogenised before analysis, then 2 g were extracted with 20 ml tetrachloroethene with ultrasonic bath. The working frequency was 40 kHz, at room temperature for 60 minutes. The extracts were purified using silica gel (0.5 g) column and eluted with a 10 ml of tetrachloroethene. The final solution were analysed with spectrophotometer-IR with Fourier Trasformed (FT-IR) at wavelength 2800-3015. The reference material used for quality control was CRM560-100-RTC. Instrument stability and response was checked using mix standard solutions. Estimated recoveries for each analyte ranged between 80% and 120%. Accuracy estimated on multiple analysis of the reference material was estimated better than ±10%. Reproducibility estimated on triplicate samples was >95%. The limit of detection was estimated as 10 mg/kg. All results were calculated with respect to dry weight.

3. Results and Discussion

3.1. Acoustic Surveys

In this section, some basic information from the acoustic surveys are reported for each investigated shipwreck: depth, length, width, height of the wrecks at the seafloor, spatial orientation along with the generated high-resolution DTM. Some of the most important physical parameters are also summarized in Table 3.

Table 3. Details of geographical setting of the studied shipwrecks

	Target 1	Target 2	Target 3	Target 4
ID	1	2	3	4
Depth (m b.s.l.)	50	35	46	180
Length (m)	75	125	31	25
Width (m)	11	25	7	10
Height from the seafloor (m)	4	15	8	4
Orientation (from N)	320°E	138°E	230°E	119°E
Final grid cell size (m)	0,35	0,25	0,35	2,5

Table 4. Mineralogy (reported in %) of the studied sediments

Shipwreck		Aragonite	Mg-calcite	Siderite	Dolomite	Quartz	Clay minerals	Kaolinite	K-feldspar
Target 1 (Kent)	a	13	58	1	-	12	15	1	-
	b	29	66	2	-	3	-	-	-
Target 2 (Pavlov V)		43	50	2	1	3	-	-	-
Target 3 (Thetis)	a	34	41	3	2	21	-	-	-
	b	17	42	3	-	23	6	-	10
Target 4 (unknown)	a	15	58	2	-	9	13	3	-
	b	14	48	2	-	9	18	3	6

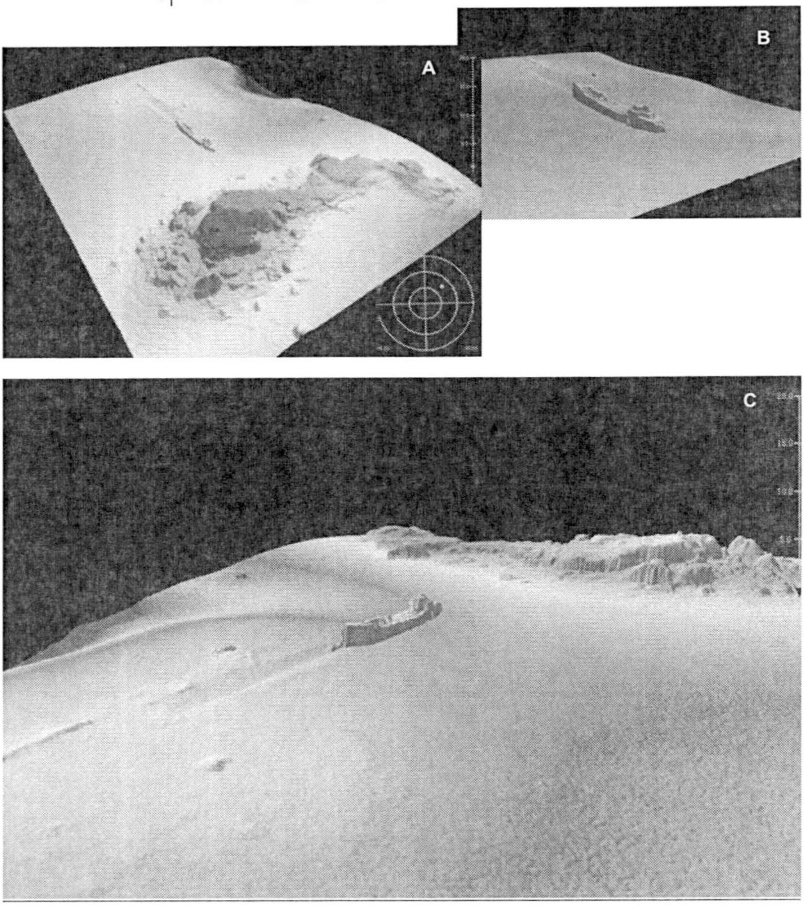

Figure 3. 3D reconstruction of the Target 1 (Kent vessel) based on the multibeam survey.

3.1.1. Target 1 (Kent Vessel)

The wreck lies at 50 m bsl depth between two seafloor highs. In the northernmost sector a rocky outcrop and some fragments of the ship are present (45 m bsl), while in the southern area a rocky seafloor high that reaches 25 m bsl occurs (Figure 3 a, b and c). A very high resolution bathymetric survey was carried out, and resulted in a DTM characterized by a 0.35 m grid cell size that shows a detailed view of the wreck and of the adjacent area. In the northernmost part of the wreck the quarterdeck and the cranes are visible, and a fore holds opening can be identified. The mooring anchor chain and a deep track (1 m) under the ship broadside indicate that the ship slipped along the slope respect to its original position. Along the way made from the wreck on the bottom, in addition to the furrow from the hull, visible part of the cargo was lost at the bottom of the sea.

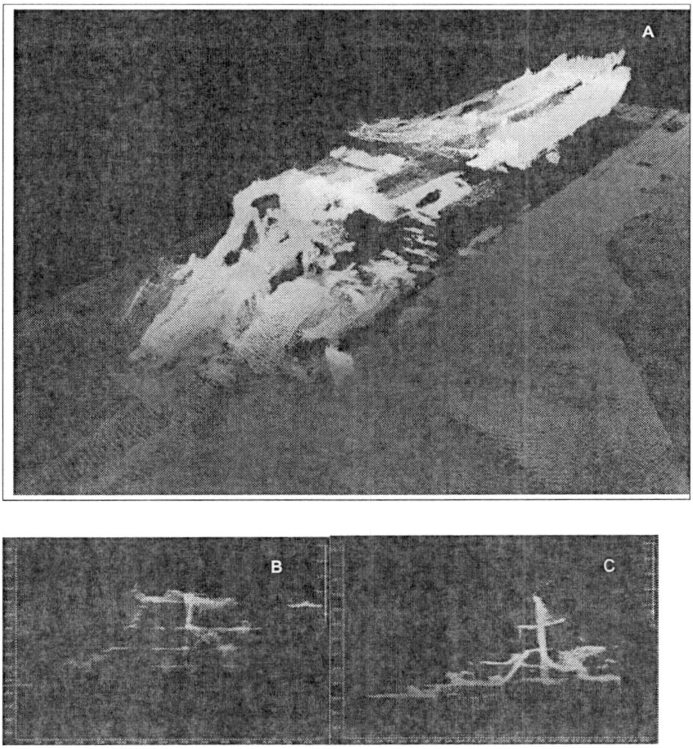

Figure 4. (a), (b) 3D reconstruction of the Target 2 (Pavlov V vessel), based on the multibeam survey,. (c) 3D image of the stern section.

3.1.2. Target 2 (Pavlov V Vessel)

MBES data results in a very high resolution DTM (0.25x0.25 m). The 3D data (Figure 4 a) allows the viewing of the broken stern (maybe due to the explosion that caused the sinking). Actually, it lies on the seabed at about 35 m depth bsl. Wreck remains are widely spread in the area, while the central part of the ship and the bow are still intacts. The shipwreck is 150 meters long and raises about 15 m from the sea-bottom. The inner bulkheads are visible through the opening in the ship hull and the decks are also recognizable (Figure 4 b and c). Some pipes run along the main deck from the bow to the stern.

3.1.3. Target 3 (Thetis Vessel)

The shipwreck lies at 46 m depth bsl, and its shallow position results in a small footprint. As a consequence, it was represented with a very high resolution grid, characterized by a 0.35x0.35 m cell size. The ship lay down on the keel, slightly bending to starboard, and a large crack is visible close to the stern. It raises about 8 m from the seafloor, in the middle section, while the stern and the bow (where the anchor's winch is visible) rise 5 m from the bottom (Figure 5).

Figure 5. 3D reconstruction of the Target 3 (Tethys vessel), based on the multibeam survey.

3.1.4. Target 4 (Unknown)

It shows a tapered shape, which is typical for shipwrecks, but it is more difficult to perfectly characterize its shape with respect to the other wrecks, because of its small size and depth (Figure 6 a). In any case, both 3D MBES images (Figure 6 a) and the SSS sonar mosaic (Figure 6 b and c) confirm its presence on the seafloor.

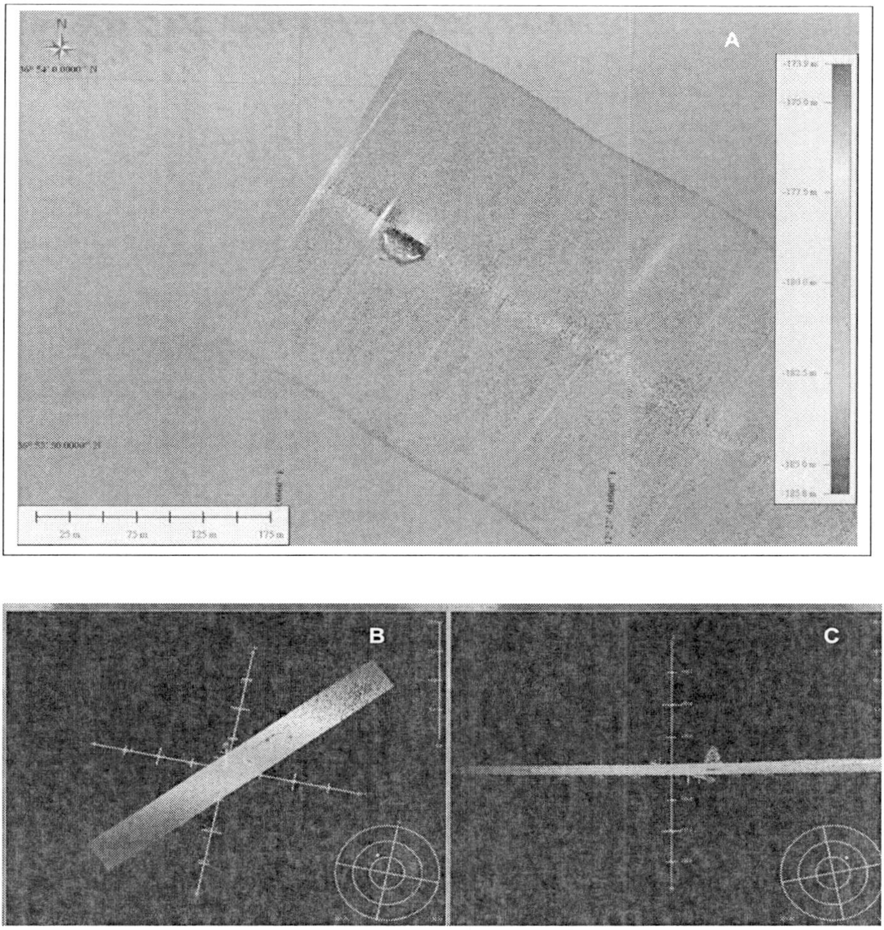

Figure 6. a) 3D reconstruction of the Target 4 (unknown vessel), based on the multibeam survey; c) Sidescan sonar data collection.

3.2. Mineralogy and Geochemistry of Sediments

3.2.1. Mineralogy

Mineralogical composition of the studied samples is summarised in Table 4. A predominance of carbonates, and in particular, Mg-calcite and aragonite occurs, with percentages between 41–66% and 13%–43%, respectively, while dolomite and siderite stay below 3%. Detrital quartz is present in variable proportion (3-23%) and results more abundant in the sample from the Target 3 area (21-23%). Finally, K-feldspar and clay minerals occur in a few samples, ranging between 6-10% and 6-18%, respectively.

3.2.2 Grain size

The sediment samples are characterized by a textural homogeneity. According to Shepard (1954) most of samples are classified as sand (86-100% sand) with the exception of sediments collected closer to the Target 1, classified as silty sands (49% sand, 31% silt, 20% clay).

3.2.3. Inorganic Geochemistry

Total concentrations of Al, Fe, As, Ba, Cd, Co, Cr, Cu, Hg, Mn, Mo, Ni, Pb, Se, Sr, V and Zn from sediment samples are reported in Table 5, including world average shales composition (Turekian and Wedepohl, 1991) here considered as natural background reference values. Most of our element contents are about 2-10 fold lower than the corresponding literature values, except for As and Cd, that display concentrations similar or higher (about twice) than those of average shales. In order to assess potential anthropic impact of the shipwrecks on the bottom sediments, some trace metals (As, Cd, Co, Cr, Cu, Ni, V and Zn) considered of relevant concern for the marine environment, are expressed in terms of enrichment factors (EF_{TM}) compared to selected reference composition values, according to the following formula:

$$EF_{TM} = (TM/Al_{sample})/(TM/Al_{reference})$$

Specifically, elements concentrations (TM) were normalized to Al in order to standardize metal concentration to changes in mineralogical composition and grain-size (Loring,1990, Salomons and Forstner, 1984). Actually, Al is usually selected as the crustal reference element for normalization, being a major crustal component and minimally affected by diagenesis (Brumsack, 1986). Subsequently, the obtained TM/Al values were compared to the average shales composition. Thus, if EF_{TM} is greater than 1, the trace metal (TM) is enriched in the samples relatively to the reference values; accordingly, if the EF_{TM} is less than 1, the element is depleted. Resulting EF_{TM} are reported in Figure 7. The degree of enrichment of metals in the samples is defined according to the reference scale reported by Sutherland (2000). The highest values are relative to As ($3<EF_{As}<125$) and Cd ($4<EF_{Cd}<25$), which result from moderately to extremely enriched in all samples, and mostly in the sediments collected close to the Target 2 ($EF_{As}=125$; $EF_{Cd}=25$) and 3 ($EF_{As}=45$; $EF_{Cd}=19$). The others elements are weakly or not enriched ($0,3<EF_{TM}<2$) in the samples, except for Fe, Cr, V and Zn in the sediments collected close to the Target 2 and 3, which show moderate/significant EF_{TM}, ranging from 3 to 10. The choice of the natural background reference is of crucial importance because it may strongly affect the conclusive evaluations. Although, most authors refer to world average shales and/or crust composition, often these two classes of values does not significantly represent the lithological/geological local features and can mislead the final interpretations. Specifically, in this work, to confirm the above metals enrichments, we assumed as local natural background levels the available pre-industrial composition determined by Tranchida et al. (2010) and Di Leonardo et al. (2012), for Sicily channel and SW Tyrrhenian sea, respectively. These reference values are reported in (Table 5) and the resulting $EF_{TM\ (local\ background)}$ are presented in Figure 8. The $EF_{TM\ (local\ background)}$ values relative to the Target 3 and Target 4 do not show evident differences with respect to $EF_{TM\ (average\ shale)}$, except for As (with $EF_{As(local\ background)}$ 2-3 times lower than $EF_{As\ (average\ shale)}$ and for Cd with a $EF_{Cd\ (local\ background)}$ about twice than $EF_{Cd\ (average\ shale)}$. These results confirm the potential pollution by As and Cd in the area surrounding the Target 3 and Target 4, but highlight a greater impact of the second metal with respect to the first one. Also, for Target 1 and Target 2, As enrichments respect to local background (low/significant) are less evident than those calculated for average shales (moderate/extremely high). Conversely, the other metals show $EF_{TM\ (local\ background)}$ 2-9 times higher than $EF_{TM\ (average\ shale)}$. In particular, Cd, Co, Cr, Cu, Ni, V, Zn increase their degree of enrichment from deficiency/moderate to moderate/very high. This primarily suggests a contamination by these elements in these two specific areas and confirm an important potential realise of As only from the Target 2.

Finally, comparisons between the $EF_{TM\ (local\ background)}$ values calculated for sediment samples collected near (a) and distal (b) to the wreck do not show significant differences, expect for Target 3, where analysed metals result

Table 5. Major, minor and trace element concentrations (mg kg^{-1}) of the studied sediments. Data from Turekian and Wedepohl (1991), Di Leonardo et al. (2012) and Tranchida et al. (2010) are also reported

Shipwreck		Al %	Fe %	As mg/kg	Ba mg/kg	Cd mg/kg	Co mg/kg	Cr mg/kg	Cu mg/kg	Hg mg/kg	Mn mg/kg	Mo mg/kg	Ni mg/kg	Pb mg/kg	Se mg/kg	Sr mg/kg	V mg/kg	Zn mg/kg
Target 1 (Kent)	*a*	1,7	0,9	7,5	38,2	0,4	2,6	20,4	6,7	<0,1	82	0,5	8,2	<0,8	<0,2	917	32,6	24,5
	b	1,1	0,6	11,9	23,8	0,2	1,6	11,9	8,2	<0,1	102	<0,1	4,3	6,9	<0,2	1428	21,6	21,6
Target 2 (Pavlov V)		0,3	1,2	56,6	9,7	0,3	1,3	10,8	1,7	<0,1	163	0,2	2,3	<0,8	<0,2	1591	44,8	21,1
Target 3 (Thetis)	*a*	0,4	1,0	27,8	24,8	0,3	1,7	6,7	2,6	<0,1	134	<0,1	2,8	<0,8	<0,2	1553	32,5	15,1
	b	1,8	1,6	24,6	61,3	0,5	3,6	24,3	3,3	<0,1	199	<0,1	8,8	<0,8	<0,2	1217	58,1	30,4
Target 4 (unknown)	*a*	2,7	2,3	12,6	53,9	0,5	6,7	32,5	6,7	<0,1	256	<0,1	13,9	<0,8	<0,2	930	55,8	37,2
	b	2,9	2,8	18,6	66,2	0,4	7,3	35,6	8,7	<0,1	280	0,4	15,5	<0,8	<0,2	1196	66,2	43,3
Average shales		8,8	4,7	13,0	580,0	0,3	19,0	90,0	45,0	-	850,0	3,0	68,0	20,0	-	-	130,0	95,0
Di Leonardo et al., 2012		4,5	-	32,8	-	0,1	2,4	13,3	2,5	-	278,6		14,6	8,4	-	-	22,7	25,5
Tranchida et al., 2010		8,1	5,1	20,2	-	0,1	15,8	76,8	25,0	-	440,7		41,2	19,7	-	-	117,2	164,7

Table 6. TOC, TN and molecular mass ratios of the analyzed samples

Shipwreck		TOC %	TN %	ΣPAHs mg/kg	ΣPCBs µg/kg	PHE/AN	FLU/PYR	AN/AN+PHE	FLU/FLU+PYR	B(a)A/B(a)A+CYR	IND/IND+B(g,h,i)I	ORIGIN
Target 1 (Kent)	*a*	1,24	0,12	1,116	0,948	4,08	1,68	0,20	0,63	0,54	0,52	Pyrolytic
	b	0,66	0,06	0,231	1,775	3,54	1,33	0,22	0,57	0,52	0,45	Pyrolytic
Target 2 (Pavlov V)		0,21	0,07	0,021	0,121	9,13	1,24	0,10	0,55	-	0,38	Pyrolytic
Target 3 (Thetis)	*a*	0,17	0,07	0,020	0,203	12,34	2,47	0,07	0,71	-	-	Pyrolytic/Petrogenic
	b	0,57	0,08	0,036	0,505	6,14	2,39	0,14	0,71	0,52	-	Pyrolytic
Target 4 (unknown)	*a*	0,45	0,08	0,240	0,208	1,45	0,42	0,41	0,29	0,49	0,67	Pyrolytic/Petrogenic
	b	0,47	0,08	0,058	0,168	0,67	1,17	0,60	0,54	0,56	0,80	Pyrolytic

clearly more enriched in the area nearest the wreck (Figure 8). This highlights the important effect that the Thetis wreck point source pollution plays on the sediments contamination.

3.2.4. Total Organic Carbon and Nitrogen

Concentrations of TOC (%) and TN (%) are listed in Table 6. TOC showed a wide range of concentrations (between 0.17 and 1.24%) and an average of 0.53%±0.3. Except for sample Target 1 (sample b) the values were lowest and/or closed to average organic content (0.68%) reported for Mediterranean sediments by Emelyanov and Shimkus (1986). The TOC/TN atomic ratio ranged from 2.46 to 10.97 displaying highest values (average 10.49) around Target 1 (samples "a" and "b"). The TOC/TN parameter is generally considered a good proxy to discriminate land versus marine origin of organic matter in coastal sediments (Sampei and Matsumoto, 2001). Generally, the proteins which are the primary nitrogen compounds of phytoplankton and zooplankton have atomic TOC/TN ratios ranging between 4 and 10. Conversely, terrestrial vascular plants and their derivatives have atomic TOC/TN ratios generally higher than 15 (Meyers, 1997). Differences in TOC/TN ratios in the studied sediments may also indicate variations in the degree of TOC preservation. During sediment diagenesis, the TOC/TN ratio can be altered by the selective degradation of organic matter components (Meyers et al., 1996). However, higher TOC concentrations, concurrent with higher TOC/TN ratios ($R=0.83$; $R^2=0.7$; $p<0.019$) in the investigated sediments, reflect improved preservation of the organic matter. TOC concentrations and TOC/TN ratios are typical of organic matter from the marine source. The presence of higher TOC concentrations and TOC/TN ratios in the surface sediments near the Target 1 could be interpreted as a more continental influence than the other sites. However, although Target 1 and 2 are both close to the coastline, the low TOC concentration and TOC/TN ratios in Target 2 imply a predominantly marine origin of organic matter, supporting the idea that other factors rather than direct land influence were the cause of the TOC and TOC/TN values detected in Target 2. One of the factors could be the geographical position of the two sites. The Target 1 is located near the Gulf of Castellamare and is likely subject to different current patterns than the Target 2. This may have produced a different accumulation of organic matter in marine sediments and a spatial/temporal variation in the degree of TOC preservation. Highest TOC and TOC/TN values in both the samples of Target 1 (point "a" and "b") demonstrate that the organic matter in sediments near and away from the vessel were likely derived from a mixing of marine and continental organic matter. It is also likely that the sediments of the Target 1 were influenced by the presence of vascular-derived carbon (i.e. seagrasses, in particular Posidonia oceanica) (Meyers et al., 1996; Twichell et al., 2002). However, mean TOC/TN values <10 in the whole analyzed samples, suggest that there is a consistent high contribution of algae (i.e. phytoplankton) to bulk of organic matter, and an increase in primary productivity due to higher nutrient input from agricultural and urban wastewater is unlikely. This hypothesis is supported also by the high relationship found between TOC and ΣPAHs of pyrolitic origin (Witt and Trost 1999).

3.2.5. Polycyclic Aromatic Hydrocarbons

Polycyclic Aromatic Hydrocarbons (PAHs) are a group of Persistent Organic Pollutants, produced by incomplete combustions of fossil fuels and urban or industrial wastes (Nriagu, 1979), and identified by the US Environmental Protection Agency (US EPA) as priority pollutants and as toxic and carcinogenic for human. Each source generates a characteristic PAHs pattern, and the PAHs distribution is governed by kinetic and/or thermodynamic properties and also by the nature of the organic matter.

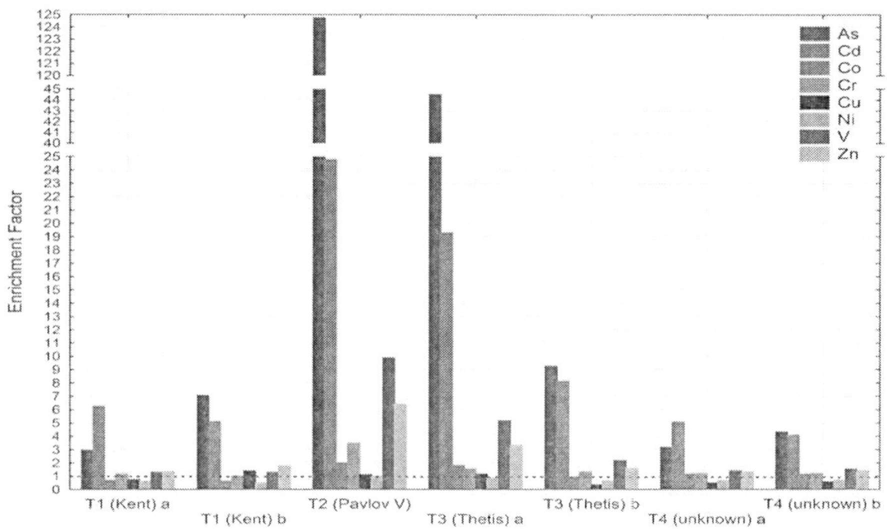

Figure 7. Estimated enrichment factors *vs.* average shale concentrations (Turekian and Wedepohl, 1991).

ΣPAHs concentration in the studied samples ranges from 0.020 to 1.116 μg kg^{-1} (Table 6). In particular, the average values for each determined PAH are: 0.053 mg/kg for PHE; 0.014 mg/kg for AN; 0.031 mg/kg for FLU; 0.020 mg/kg for PYR; 0.025 mg/kg B(a)A; 0.018 mg/kg for CRY; 0.028 mg/kg for B(b)F; 0.023 mg/kg for B(k+j)F; 0.013 mg/kg for B(a)P; 0.018 mg/kg for B(e)P; 0.004 mg/kg for PER; 0.022 mg/kg for IND; 0.005 mg/kg for D(a,h)A and 0.014 mg/kg for B(g,h,i)P. The highest values of ΣPAHs were found in both sediments near to the Target 1 and 4. A number of PAH isomeric ratios have been routinely used to identify different sources that contribute PAHs to environmental samples (Yunker et al., 2002; Budzinski et al., 1997; Baumard et al., 1998). The interest of these indices relies on the fact that during low-temperature processes, such as those occurring during catagenesis of organic matter leading to the formation of petroleum, the PAHs distribution is governed by thermodynamic properties.

On the contrary, for high-temperature processes, such as pyrolysis of organic matter, their distribution is rather governed by kinetic characteristics (Baumard et al. 1998).

In particular, environmental sources of PAHs include petroleum, sewage, biomass combustion (grass, wood, or coal combustion), or petroleum combustion and are generally derived from anthropogenic sources. We illustrate below the considerations gathered from the examination of PAH molecular ratios used in order to estimate the origin of PAHs for each investigated site. (Sicre et al., 1987; Budzinski et al., 1997; Baumard et al., 1998; Soclo et al., 2000; Magi et al., 2002) (Table 6). PHE/AN values lower than 10 and FLU/PYR values lower than 1, indicate contamination of PAHs due to combustion processes (pyrogenic origin). AN/AN+PHE ratio was used to distinguish between combustion and petroleum sources. In detail, AN/AN+PHE lower than 0.10 indicates petroleum impact (petrogenic origin), while values higher 0.10 indicate a dominance of petroleum combustion. B(a)A/B(a)A+CRY values above 0.35 is considered indicative of combustion, instead, values between 0.20 and 0.35 indicate either petroleum or petroleum combustion, and values lower than 0.20 imply petroleum sources (Botsou and Hatzianestis 2012). FLU/FLU+PYR values between 0.4 and 0.5 are characteristic of combustion of liquid fossil fuels, whereas values higher than 0.5 indicate grass, wood or coal combustion (Yunker et al. 2002). IND/IND+B(ghi)P values lower than 0.20 suggest petroleum as probable source. Ratio values higher than 0.50 imply grass, wood or coal combustion.

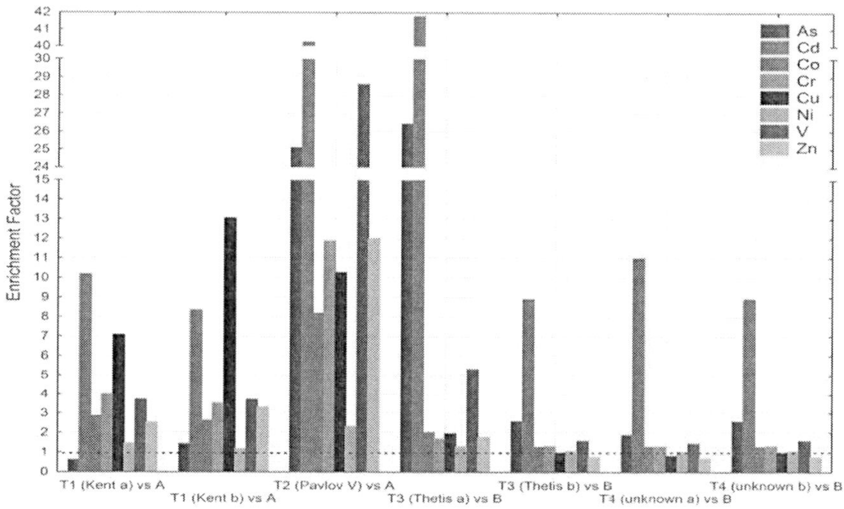

Figure 8. Estimated enrichment factors *vs.* estimated local background (a) Di Leonardo et al. (2012) and (b) Tranchida et al. (2010).

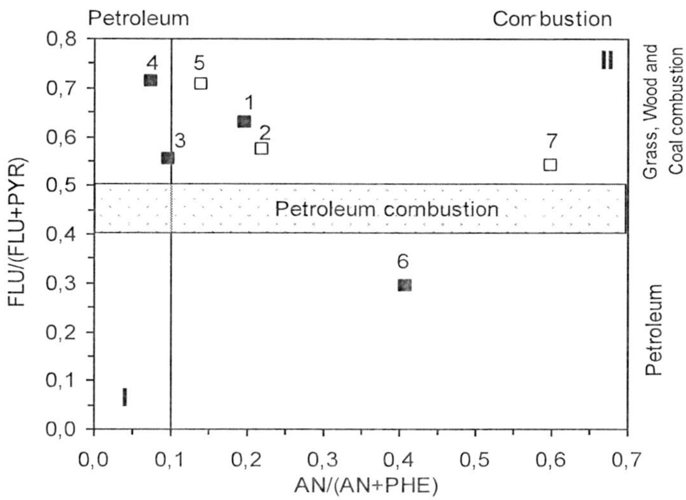

Figure 9. PAHs cross plot of molecular ratio of FLU/(PYR+FLU) vs. AN/(AN+PHE). Lines represent the reported values for PAHs derived from petrogenic (I) or pyrolytic sources (II). (1) and (2) represent Target 1; (3) represents Target 2; (4) and (5) represent Target 3; (7) represents Target 4.

In this study a predominance of pyrolytic origin was established for PAHs recorded in the studied sediments, except for two samples (Figure 9). In proximity of Target 3 and 4, PAHs formation is probably related to both pyrolytic and petrogenic processes. A petrogenic component established for samples near to the Target 3 and 4 could suggest an incomplete microbial degradation of PAHs, which instead probably occurred in the other samples. To corroborate this hypothesis we know from historical chronicles that Target 3 and 4 are more recent shipwrecks compared to Target 1 and 2.

3.2.6. Polychlorinated Biphenyls

Polychlorinated biphenyls enter the aquatic environment through a variety of pathways including atmospheric transport and deposition, river inputs, industrial and municipal wastewater discharges. Due to their low solubility in water and high octanol-water partition coefficients, PCBs have a strong affinity for particulate matters and ultimately accumulate in bottom sediments. The use of PCBs as raw material or chemical intermediate was banned in the EU in 1985 (council Directive 85/467/EEC 1985). Thus, in terms of aquatic environmental monitoring, the concentration of PCBs in surface sediments provides information on recent contamination. ΣPCBs in the studied samples vary from 0.168 to 1.775 $\mu g\ kg^{-1}$.

Once again, the highest ΣPCBs values were determined in sediments close to the Target 1. In particular, penta-PCBs (101 and 114) and hexa-PCBs (138 and 153) are the most abundant congeners (Table 6). This basically suggests a potential contribution and leakage of fluids from the electrical transformers and electrical devices present on board of the vessel.

3.2.7. Organo-chlorine Pesticides

Organo-chlorine pesticides (OCPs) are a class of organic compounds commonly used for pest control in agriculture. They result as a potential health risk due to their bioaccumulation and potential toxicity to humans and wildlife. Furthermore, DDT and DDE are known endocrine disrupting compounds (Amaral 2002). It is known that DDT degradation in soil occurs by both biotic and abiotic processes to form DDD and DDE (Foght et al., 2001).

In the studied samples, all the OCPs are below the limit of quantification, except for sample "b" of the Target 1, where DDE, DDD and DDT are present with relatively high values of 59.6, 1292.1 and 2475.6 μg/kg, respectively. Historical information report that the Kent vessel transported a load of phosphates, generally containing significant concentration of OCPs. This could explain effects of loss and leakage of these contaminants in the surrounding environment.

Finally, concentrations of DRO are ever below the limit of quantification (< 50 mg/kg).

CONCLUSION

The integrated acoustic and organic/inorganic geochemical dataset collected around the four shipwrecks which occur at the bottom of the central Mediterranean sea offered a good opportunity to investigate contamination effects on the surrounding seabed and evaluate potential environmental risk. Evidence of different leakages and/or transfer of hazardous contaminants from the four investigated abandoned vessels to the sediments, in terms of the analysed heavy metals (specifically As and Cd) and priority organic compounds (some congeners of PCBs and PAHs), suggest an urgent need of better understanding the crucial role played by this potential sources of pollutants to the marine environment and ecosystem. Moreover, the detailed modes of distribution and dispersion of contaminants in the area where shipwrecks occur, although here explored with a reduced number of samples, show a highly site specific behaviour that needs accurate sampling strategy and data collection. The fate of hazardous compounds on the seabed and marine ecosystem is strongly related to origin and source of contaminants, but also to time of exposition.

An international effort is mandatory for a most appropriate exploration and registration of such a kind of spot anthropic impact along with appropriate and complete exploration of chemical contamination of the different marine compartments, including living resources. Finally, we think that a suitable ecosystem approach to monitor this pollutant point sources is prodromal and mandatory for an appropriate effort of remediation and/or mitigation of the potential negative effects on a productive and healthy ocean.

ACKNOWLEDGMENTS

We thank all the crew of the R/V Urania for their valuable professional job support during the cruises. Constructive reviews by an anonymous reviewer helped improve the manuscript.

References

Amaral Mendes J.J. 2002. "The endocrine disrupters: a major medical challenge". *Food Chem. Toxicol.* 40, 781-788.

Barahona, E., Huertas, F., Pozzuoli, A., Linares, J., 1982. Mineralogia e genesi dei sedimenti della provincia di Granata (Spagna). *Miner. Petrogr. Acta,* 26, 61–90.

Baumard P, Budzinski H, Michon Q, Garrigues P, Burgeot T, Bellocq J (1998) Origin and bioavailability of PAHs in the Mediterranean Sea from mussel and sediment records. *Estuarine Coastal Shelf Sci* 47:77–90.

Botsou F., Hatzianestis I. Polycyclic aromatic hydrocarbons (PAHs) in marine sediments of the Hellenic coastal zone, eastern Mediterranean: levels, sources and toxicological significance. *J Soils Sediments* (2012) 12:265–277, DOI 10.1007/s11368-011-0453-1.

Brumsack H. J. (1986) The inorganic geochemistry of Cretaceous black shales (DSDP Leg 41) in comparison to modern upwelling sediments from the Gulf of California, 447-462. In *Geological Society,* London, Special Publications 21 (1).

Budzinski H, Jones I, Bellocq J, Piérard C, Garrigues P (1997) Evaluation of sediment contamination by polycyclic aromatic hydrocarbons in the Gironde estuary. *Mar Chem* 58:85–97.

Di Leonardo R., Cundy A.B., Bellanca A., Mazzola A., Vizzini S. Biogeochemical evaluation of historical sediment contamination in the Gulf of Palermo (NW Sicily): Analysis of pseudo-trace elements and stable isotope signals. *Journal of Marine Systems* 94 (2012) 185–196.

EC (1985) Council Directive 85/467/EEC of 1 October 1985 amending for the sixth time (PCBs/PCTs) Directive 76/769/EEC on the approximation of the laws, regulations and administrative provisions of the Member States relating to restrictions on the marketing and use of certain dangerous substances and preparations.

EC (2008) Directive 2008/56/EC of the European Parliament and of the Council of 17th June 2008 establishing a framework for community action in the field of marine environmental policy (Marine Strategy Framework Directive). *Official Journal of the European Union L* 164: 19–36.

Emelyanov, E.M., Shimkus, K.M., 1986. Geochemistry and sedimentology of the Mediterranean sea. In: Evans, G., Bush, P.R. (Eds.), *Sedimentology and petroleum geology.* Imperial College, D. Reidel Publishing Company, London, p. 553.

Foght J., April T., Biggar K., Aislabie J. Bioremediation of DDT-Contaminated Soils: A Review. *Bioremediation Journal,* vol. 5, no. 3, pp. 225-246, 2001, DOI: 10.1080/ 20018891079302.

IHO (2008) *International Hydrographic Organization - IHO Standards for Hydrographic Surveys,* 5th Edition, february 2008 - Special Publication N° 44.

Loring, D.H., 1990. Lithium—a new approach for the granulometrical normalization of trace metal data. *Marine Chemistry* 29, 156–168.

Magi E, Bianco R, Ianni C, Di Carro M (2002) Distribution of polycyclic aromatic hydrocarbons in the sediments of the Adriatic Sea. *Environ Pollut* 119:91–98.

Meyers, P.A., 1997. Organic geochemical proxies of paleoceanographic, palaeolimnological, and palaeoclimatic processes. *Organic Geochemistry,* 27; 213-250.

Nieuwenhuize J, Maas YEM. MIddelburg JJ (1994) Rapid analysis of organic carbon and nitrogen in particulate materials. *Mar Chem* 45, 217-224.

Nriagu JO. 1979. Global inventory of natural and anthropogenic emission of trace metals to the atmosphere. *Nature,* 279: 409-411.

OSPAR (2008a) *Code of Conduct for Responsible Marine Research in the Deep Seasand High Seas of the OSPAR Maritime Area.* Available: http://www.ospar.org/v_

OSPAR (2008b) *Summary Record of the Meeting of the OSPAR Commission,* Brest (France), 23–27 June 2008, OSPAR 08/24/1-E. Available: http://www.ospar.org. Accessed 19 August 2009.

Salomons, W., Forstner, U., 1984. *Metals in the Hydrocycle.* Springer, Berlin.

Sampei, Y. and Matsumoto, E. 2001. C/N ratios in a sediment core from Nakaumi Lagoon, southwest Japan – usefulness as an organic source indicator. *Geochemical Journal,* 35: 189-205.

Schubert, C. J. and Nielsen, B. (2000) Effects of decarbonation treatment on $\delta^{13}C$ in marine sediments. *Mar. Chem.* 72, 55–59.

Schultz, L.G., 1964. Quantitative interpretations of mineralogic composition from X-ray and chemical data for the Pierre Shale. *US Geol. Surv. Prof. Pap.,* 391-C, C1–C31.

Sicre, M.A., Marty, J.C., Saliot, A., Aparicio, X., Grimalt, J., Albaiges, J., 1987. Aliphatic and aromatic hydrocarbons in different sized aerosols over the Mediterranean Sea: occurrence and origin. *Atmos. Environ.* 21, 2247–2259.

Soclo HH, Garrigues P, Ewald M (2000) Origin of polycyclic aromatic hydrocarbons (PAHs) in coastal marine sediments: case studies in

Cotonou (Benin) and Aquitaine (France) areas. *Mar Pollut Bull* 40:387–396.

Standard Operating Procedure (2007) *Instrumental analysis for total organic carbon and total nitrogen in sediments for the USGS reston, virginia environmental organic geochemistry laboratory.* http://pubs.usgs.gov/of/2007/1362/Appendix%206.pdf.

Sutherland, R.A., 2000. Bed sediment-associated trace metals in an urban stream, Oahu, Hawaii. *Environmental Geology* 39, 611–627.

Tranchida, G., Bellanca, A., Angelone, M., Bonanno, A., Langone, L., Mazzola, S., Neri, R., Patti, B., 2010. Chronological records of metal deposition in sediments from the Strait of Sicily, central Mediterranean: Assessing natural fluxes and anthropogenicalteration. *J. Mar. Syst.* 79, 157–172.

Turekian, K., Wedepohl, K., 1991. Distribution of elements in some major units of the earth's crust. *Geol. Soc. Am. Bull.,* 72, 175-192.

UN (1992) *Convention on Biological Diversity. United Nations Environment Programme.* Available: http://www.cbd.int/convention/convention.shtml. Accessed 21 August 2009.

UN (1995) *Agreement for the implementation of the provisions of the United Nations Convention on the Law of the Sea of 10 December 1982 relating to the conservation and management of straddling fish stocks and highly migratory fish stocks.:* United Nations Available: http://daccess-ddy.un.org/doc/UNDOC/GEN/N95/274/67/PDF/N9527467. pdf?OpenElement. Accessed 21 August 2009.

Verardo DJ, Froelich IN, McIntyre A (1990) Determination of organic carbon and nitrogen in marine sediments using the Carlo Erba NA-1500 analyzer. *Deep-Sea Research* 37, 157–165.

Yunker M. B., McDonald, R.W., Vingarzan, R.,Mitchell, R. H., Goyotte, D., Sylvestre, S., 2002. PAHs in the Fraser River basin: a critical appraisal of PAH ratios as indicators of PAH source and composition. *Organic Geochemistry* 33, 489–515.

In: Mediterranean Sea
Editor: Terrence B. Hughes

ISBN: 978-1-62618-238-7

Chapter 3

DISPOSAL OF PRODUCED FORMATION WATER FROM OFFSHORE GAS PLATFORMS IN THE MEDITERRANEAN SEA: A PARAMETRIC STUDY ON DISCHARGE CONDITIONS AIMED AT MITIGATING RISKS FOR THE MARINE ENVIRONMENT

Daniela Cianelli[1,2], Rossella Di Mento[1], Loredana Manfra[1], Federico Oteri[1], Anna Maria Cicero[1] and Enrico Zambianchi[2*]

[1]ISPRA, Advanced Institute for Environmental Protection and Research, Via Vitaliano Brancati, Roma, Italy

[2]Department of Environmental Sciences, "Parthenope" University of Naples, Centro Direzionale Isola, Napoli, Italy

ABSTRACT

The offshore oil and gas production in the Mediterranean Sea has considerably grown over the last 20 years, introducing a wide range of potential risk factors for the health of marine ecosystems. Crucial impacts may derive from discharges of produced formation water (PFW), a by-

* Corresponding author enrico.zambianchi@uniparthenope.it.

product of gas extraction, which to date represents approximately 80% of the waste and residual yield from natural gas production activities. The disposal of PFW may be carried out through the reinjection into the reservoir, the transport onshore or the discharge into seawater. We focused our attention on PFW discharged by submerged outfalls in the Adriatic Sea (a shallow Mediterranean sub-basin characterised by an intense extraction and production activity).

The effects of PFW discharge on the marine environment depend on the characteristics of the effluent as well as on the dispersion processes occurring in the receiving ambient fluid. In order to assess these potential environmental impacts we apply a numerical dispersion model integrated with a monitoring plan developed for the Adriatic Sea.

We simulated the initial mixing of PFW discharged from a selected offshore gas platform during the summer season, and used field data (samples of ambient sea water, current intensity and direction measurements, temperature and conductivity profiles) collected in the vicinity of the platforms from 2008 to 2010 during monitoring surveys. We evaluated several different discharge scenarios (in terms of environmental conditions and effluent characteristics) and three working hypotheses for the discharge operations. The results of this parametric study allowed us to identify some sort of "optimal discharge practices" for PFW in the Adriatic Sea, which should promote the rapid dilution of the effluent and mitigate the impacts of the PFW on marine ecosystems.

INTRODUCTION

The anthropic activities and the continuous input of large amounts of pollutants into the marine environment are potential risk factors for the quality of coastal waters and the health of marine ecosystems. The worldwide offshore oil and gas production has grown considerably over the last several years (Wilson III and Heath, 2008; BP, 2011; Enerdata, 2011); this is definitely true for the Adriatic Sea (Mediterranean Sea), a basin characterised by a shallow and wide shelf. The wide range of potential impacts of offshore oil and gas activities requires specific management measures. In particular crucial impacts may derive from offshore installations that physically disturb seabed habitats, which cause atmospheric emissions, noise, discharges of cuttings, piles and discharges of produced formation water (PFW) (OSPAR, 2009; Manfra et al, 2010; Manfra and Maggi, 2012). This latter represents a key issue due to the origin, characterisation and disposal of PFW.

The PFW is a by-product of oil and gas extraction, which to date represents the largest volume waste from the production phase of the offshore oil and gas facilities, constituting approximately 80% of the wastes and residuals yield from natural oil and gas production operations (McCormack et al., 2001; Berry, 2005). The PFW is usually a saline water that may contain a number of potential contaminants such as heavy metals, hydrocarbons, phenols, ammonia, radionuclides and chemicals (demulsifiers and corrosion inhibitors) added at times during the operation of the well (OGP, 2005; Scott et al., 2007). The concentrations of these compounds are typically high when compared to the receiving environment (Manfra et al., 2007).

The disposal of PFW is carried out by reinjection into the reservoir, onshore transport, or the discharge into the ocean. We focus our study on the third way of disposing them: releasing them into the marine environment by means of submerged outfalls.

After the release into the sea, PFW undergoes several different processes such as dilution into the ambient fluid, evaporation and biodegradation, volatilization towards the atmosphere and settling at the bottom. Our research has been aimed at investigating a crucial process determining the dilution of PFW in the seawater, namely the mixing of the effluent with the ambient fluid (e.g. Baumgartner et al., 1994). Such a dispersion process occurs in two phases: a rapid initial (near-field) mixing phase taking place immediately after the release within the first tens of meters from the discharge point, and the subsequent far-field passive dispersal phase that evolves at larger distance over time scales of hours or days.

The impact of PFW discharges on the marine environment strictly depends on the characteristics of the effluent and the receiving ambient fluid. In order to assess the potential effects and risks for marine ecosystems it is of crucial importance to schedule monitoring plans. Monitoring programs have been developed worldwide (e.g. North Sea, Gulf of Mexico, Adriatic Sea) characterised by an intense extraction and production activity. These monitoring programs are mainly aimed at: establishing the environmental background level before the PFW discharge; identifying spatial and temporal evolution of physical, chemical and biological parameters during the PFW discharge; assessing mitigation measures and defining guidelines supporting regulations and decisions. The complex dispersion processes, which determine the fate of the effluents discharged into the marine environment, makes it difficult to assess the behavior of the PFW with only field observations (Cianelli et al., 2011). Our work stems from the observation that these field measurements are often unpractical and expensive, thus an exhaustive assessment of the effect of PFW discharge on marine environment needs to integrate, in a monitoring plan, in situ sampling with numerical modelling of effluent dispersion.

In a previous work (Cianelli et al., 2008a), applying an integrated numerical-chemical approach in the Adriatic Sea, we reconstructed the temporal and spatial distribution of the PFW plume discharged from three offshore gas platforms and identified the main environmental factors affecting plume behavior. These results also indicated that the rapid initial dispersion times might cause negligible or non-toxic effects on marine organisms.

The research activity we present here starts from those results and is aimed at assessing the role of environmental and discharge conditions in promoting a rapid dilution of produced water.

Applying the integrated observational - numerical monitoring protocol that we put together, we developed a parametric study on one selected platform evaluating several different discharge scenarios (dispersion under various stratification conditions and effluent characteristics).

We here apply a Lagrangian numerical model simulating the initial mixing of PFW discharged from a selected offshore gas platform during the summer season, and use field data (marine water samples, current intensity and direction, temperature and conductivity profiles) collected close to the platforms at increasing distances from the discharge source. In order to assess the influence of both geometrical diffuser parameters and effluent flows on the near field dilution we set three working hypotheses for the discharge operations. Three values typical of coastal zones have also been chosen for the horizontal diffusivity parameter acting in the far-field zone, each one representing low, high and intermediate energy dissipation conditions. The results of this work suggest some "optimal discharge practices" for PFW in the Adriatic Sea, which could most effectively promote the rapid dilution of the effluent and mitigate the impacts of the PFW on the marine environment.

Study Area

The Mediterranean Sea holds approximately 0.4% of world reserves of oil and gas (Pinder, 2001), which in the last twenty years has originated a great expansion in offshore activities in this basin. The offshore structures (about 90 platforms) for the oil and gas industry are mainly located in the Northern and Central Adriatic Sea (Maggi et al., 2007; Trabucco et al., 2012), which occupies the northernmost part of the Mediterranean basin (Figure 1).

The Adriatic Sea is located between Italy and the Balkan Peninsula and is linked with the Ionian Sea (Mediterranean Sea) in its southern part through the Strait of Otranto. The Adriatic Sea is a semi-enclosed basin landlocked on the northern, eastern and western part; it is characterised by a wide shelf reaching a maximum depth of ~ 100 m in its northernmost section, and showing an average depth less than 200 in its middle portion; the southern half of the Adriatic presents a maximum depth of about 1200 m in the center.

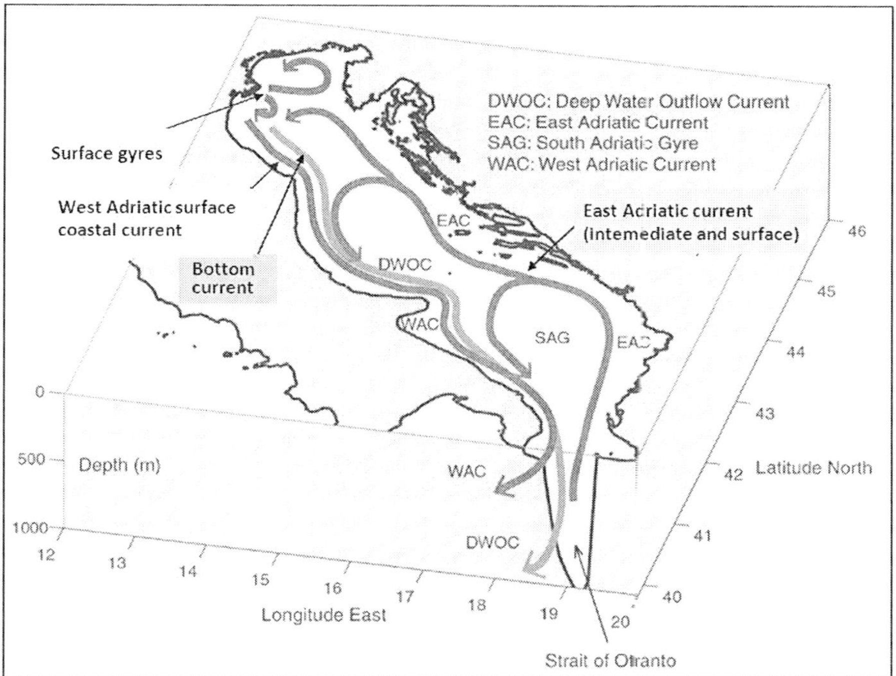

Figure 1. Adriatic Sea thermoaline circulation adapted from Cushman-Roisin, 2001.

The hydrology and meteorology of the basin are highly influenced by its position enclosed by land; as a result, the circulation and the distribution of Adriatic seawater masses strongly depend on the characteristics of air-sea and lateral (river runoff) fluxes (e.g. Cushman-Roisin, 2001).

The water exchange between the Adriatic and the Mediterranean Sea occurs through the Strait of Otranto, with an inflow along the eastern coast and an outflow along the western boundary (e.g., Russo and Artegiani, 1996).

The Adriatic circulation shows a relevant seasonal and inter-annual variability in response to changes in the heat flux and wind regimes. The winter circulation shows a prevalence of Mediterranean inflow mainly due to southerly winds, while the summer season is characterised by a slightly stronger outflow of fresh and Adriatic water (e.g. Cushman-Roisin, 2001). Both the horizontal and vertical circulations play a crucial role in the transport and dispersion of marine pollutants and PFW discharged in the Adriatic Sea.

Experimental Monitoring Strategy of the Adriatic PFW Discharge

The rapid spatio-temporal variability of the oceanographic conditions may directly influence the vertical displacement of the effluent within the water column as well as the near field dilution process. In order to evaluate the fate and dispersion of the PFW discharged into the Adriatic Sea, an Adriatic Platform Monitoring (APM) plan has been expressly conceived by ISPRA (Trabucco et al, 2012: Di Mento et al., 2009).

The monitoring of the Adriatic PFW discharge includes the chemical-physical analyses of water and sediment samples, together with bioaccumulation investigations. Moreover the temperature and salinity vertical structure of the water column is monitored by means of CTD profiles collected with a multi-parametric probe equipped with additional sensors for fluorescence, transmittance, dissolved oxygen and pH. Vertical profiles of currents are measured by an Acoustic Doppler Current Profiler (ADCP).

In order to determine chemical concentrations of nutrients, BTEX and total petroleum hydrocarbon in the seawater, samples are collected by means of Niskin bottles, on a transect starting from the point of discharge and following the direction of the in situ measured current. To assess the presence of contaminants in the sediments, samples are taken by Van Veen grab or box corer, on a transect oriented in the direction of the dominant current. Mussels are collected at two different depths (surface and -12 m), on the leg nearest to the discharge point, along the main current, to investigate accumulation of contaminants in their tissues.

The monitoring plan implies one or two sampling surveys before the beginning of the platform operation, two cruises after the beginning of the discharge and then one survey for each following year of activity.

INITIAL DILUTION MODEL

Since the 1980's numerical dilution models have been applied worldwide to several environmental pollution problems such as the disposal of industrial wastes and drilling fluids from offshore oil and gas exploration, and the discharge of contaminants in rivers and lakes.

The initial dilution of the effluent occurring in the near field zone is mainly due to buoyancy and to momentum associated with the discharge itself. The mixing process by which the ambient fluid incorporates the effluent is known as entrainment and the dilution process can be assumed as a by-product of entrainment. In order to simulate the near field mixing of the PFW discharged from the selected gas platform we here applied UM3 (Three-dimensional Updated Merge), a mathematical model commonly used by US EPA for marine discharges and available in the interface Visual Plumes (e.g. Frick et al., 2004). UM3 is a three-dimensional Lagrangian model that accurately quantifies the entrainment computing both the Taylor entrainment and the projected-area-entrainment (PAE).

The Taylor entrainment is due to the shear between the effluents discharged and the receiving water, while the PAE is the rate at which mass is incorporated into the plume in the presence of an ambient current. Both these processes determine the growth of the plume element and its dynamics (Baumgartner et al., 1994) on the basis of the entrainment equation (e.g., Baumgartner et al., 1994) which computes the increment of the mass *dm* entrained in the time step *dt* as the sum of the forced and the Taylor entrainment terms:

$$\frac{dm}{dt} = -\rho A_P \overline{U} + \rho A_T v_T \qquad (1)$$

The forced term is given by the product of the ambient density ρ, the projected area A_P and the average velocity of ambient fluid through the projected area $\overline{U}$. The Taylor term is the product of the ambient density ρ, the area of the plume element in contact with ambient fluid A_T and the Taylor aspiration speed v_T.

UM3 also computes the equations for conservation of mass, momentum and energy at each time step along the plume trajectory.

The entrainment process tends to decrease the density difference between the effluent and the medium thus favoring the initial dilution. To quantify the

PFW dilution UM3 computes a commonly used parameter, the average dilution factor S_a, which is the ratio between the effluent concentration and the cross-sectional average concentration in the plume.

The densimetric Froude number is a similarity parameter, which provides an estimate of the relative importance of momentum and buoyancy in driving the initial mixing and the plume rise. A plume dominated by buoyancy is characterised by Froude number less than 1, while values between 10 and 100 indicate a dominated momentum plume (e.g. Frick et al., 2002).

In the present work the densimetric Froude number is computed as:

$$Fr = \frac{U_0}{\sqrt{gD\frac{\Delta\rho}{\rho_e}}} \tag{2}$$

where D is the outfall pipe port diameter, $\Delta\rho$ is the density difference between the effluent and the ambient water at the depth of the port, ρ_e the effluent density and g is the gravity. U_0 represents the effluent outflow velocity, given by:

$$U_0 = \frac{4q}{\pi \cdot D^2} \tag{3}$$

with q is the volume flux at outfall pipe port.

In order to characterize the stability of the water column at discharge depth we compute for our platform the Brunt-Väisälä frequency N:

$$N = \left[-\frac{g}{\rho_{ref}} \frac{\partial \rho}{\partial z} \right]^{\frac{1}{2}} \tag{4}$$

where ρ_{ref} represents the ambient fluid density at the depth of the discharge and $\partial\rho/\partial z$ is the vertical density gradient in correspondence of the outfall depth. N has been computed by finite differencing the measured density profile in a vertical range of 2 meters centered around the depth of discharge.

As initial conditions of UM3 we used the measured current data, and density profiles of the receiving water column as well as the measured PFW concentrations and the outfall pipe geometrical features. This model allowed us to consider the diameter, the orientation of the port and the depth of discharge as further outfall features, which may affect the effluent velocity and the plume trajectory. The numerical output provides the following wastefield characteristics at the end of the near field:

- the average dilution factor
- the plume rise height
- the plume thickness and diameter
- the length of the near field
- the initial mixing time

In order to simulate the PFW plume performance under a range of environmental and discharge conditions, we need to initialize the model with several field data collected during the monitoring surveys; UM3 addresses this issue allowing the use of time series data as input parameters.

Ambient properties such as currents intensity and stratification conditions also affect the entrainment and the rate of dilution thus determining the concentration of contaminants in the receiving waters. A low ambient current will limit the initial dilution process; in this case the plume will extend down current reaching a maximum distance equal to the water depth (Roberts et al., 1989).

On the other hand, intense currents can increase the shear between the plume and the medium, thus enhancing turbulence, which in turn favors the entrainment of marine water within the effluent, and consequently its dilution.

The ambient salinity and temperature influence the plume buoyancy determining the equilibrium density of the plume in the water column. If the ambient fluid has constant density, the density of the diluted effluent will asymptotically reach the environmental density without equalling it, thus continuing its rise (or sink in the case of negative buoyancy), possibly reaching the surface (or the bottom). Worth noticing is that, moreover, on the basis of previous dilution model comparisons (e.g. Roberts and Tian, 2004) UM3 has proven to provide the best simulation of the initial dilution and length of near field zone.

FIELD DATA

In this work we selected a platform located in the central-northern Adriatic Sea at a 56 km distance from the Italian coast. The bottom depth in correspondence to the platform is 70 m and the PFW is discharged from a single port outfall pipe oriented downward at 36 m depth. The field data utilized in this study come from three PFW chemical characterizations and three water column surveys carried out from 2008 to 2010 during the summer season.

The monitoring surveys were carried out during the PFW discharge on a transect displayed along the ambient current direction as measured at the depth of the discharge point. Seven stations were located at increasing distance from the discharge source (0 m – close to the platform – 7 m, 15 m, 25 m, 50 m, 100 m and 500 m) (Figure 2).

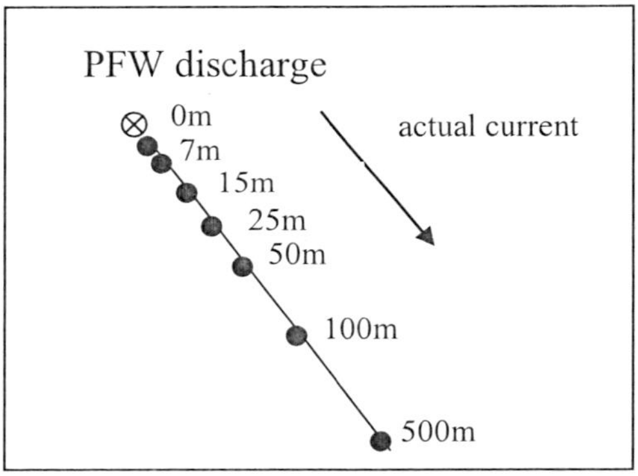

Figure 2. Design of water sampling stations near the PFW discharge point for hydrological parameters (figure not to scale).

The CTD profiles were collected using a SBE 25 sonde (Sea-Bird Electronics, WA, USA) and the current profiles were recorded continuously by a ship-mounted 307 kHz WorkHorse Broadband ADCP (RD Instrument, CA, USA). The bottom-tracking method was used to calculate the absolute water velocity by subtracting the boat velocity from the relative one measured by the ADCP. The WinADCP Software from RD Instruments was used to transform the raw ADCP data for further post-processing with Matlab routines.

RESULTS

The PFW volumes discharged from 2007 to 2010 and the environmental field data measured during the summer monitoring surveys of 2008, 2009 and 2010 have been used with the diffuser geometrical parameters to simulate 72 discharge scenarios from the selected platform. In all the numerical simulations of our study we considered the PFW as a conservative tracer, thus we did not account for any loss of material contained in the PFW through removal or degradation processes; therefore our results may be considered a worst case of PFW discharge.

The selected platform is located close to the coast (~ 50 km) where the maximum depth of the water column is 70 m, while the discharge depth is 36 m (see Table 1). In order to evaluate the effect of the geometrical diffuser features and of the effluent flows on the PFW dilution we considered a (downward) vertical oriented pipe and adopted two values for the diffuser diameter: the theoretical value of 45 cm and the real one of 72 cm. The annual PFW volumes released from 2007 to 2010 have been scaled hypothesizing three discharge conditions (see Table 1 for specific values):

fv_1 – intermediate effluent flow: throughout the year the effluent is released once a week continuously over 24 hours;
fv_2 – maximum effluent flow: throughout the year the effluent is released once a day continuously over 3 hours;
fv_3 – minimum effluent flow: throughout the year the effluent is released three times a week continuously over 24 hours.

We also ran simulations using three values for the horizontal diffusion coefficient (see Table 1 for specific values) typical of coastal areas:

- kd_1 represents low turbulence conditions = 0.0001 m^2/s;
- kd_2 is associated to high energy dissipation conditions = 0.0005 m^2/s;
- kd_3 is an intermediate value between kd_1 and kd_2 = 0.0003 m^2/s;

lower values of kd obviously provide more conservative estimates of far-field dilution (Frick et al, 2002).

The effluent features are represented by the PFW salinity, temperature and flow; the values of T_{PFW} and S_{PFW} are kept constant for all the simulated scenarios.

The duration of simulations has been fixed at 9h, which reproduces the temporal scale of the near field dispersion processes; moreover UM3 allows for an a-priori determination of the acute and chronic mixing zone limits, which have been set at 100 m and 1000 m respectively.

Table 1. Input model parameters

Depth (m)	Discharge depth (m)	Diffuser diameter (cm)	Discharged volumes (mc/a)	T_{PFW} (°C)	S_{PFW} (psu)	Hydrographic Data (CTD) (T, S)	Current data ADCP
70	36	45	2007 14879	23,4	38,93	2008	2008
		72	2008 13181			2009	2009
			2009 9147			2010	2010
			2010 5603				

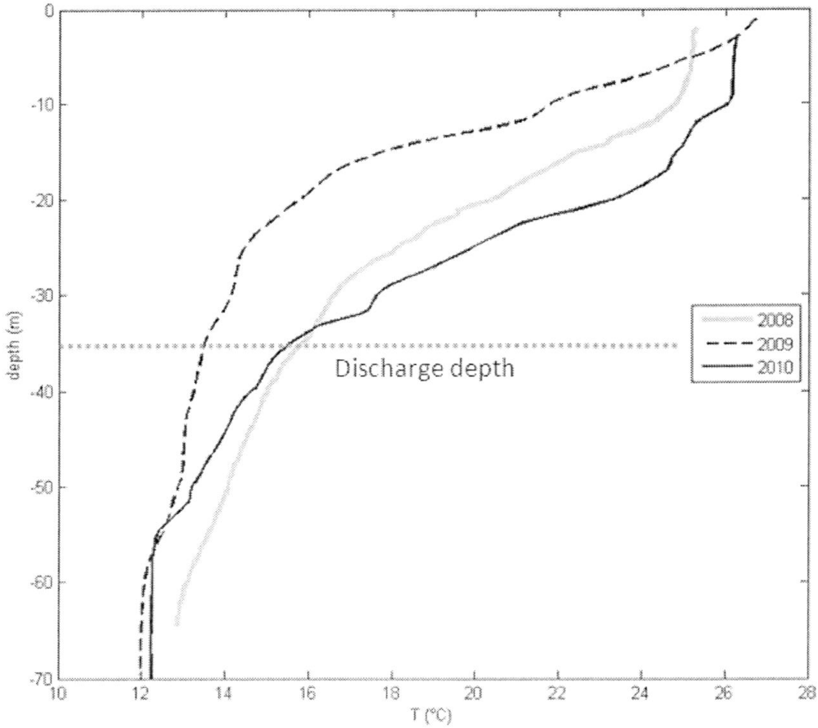

Figure 3. (Continued).

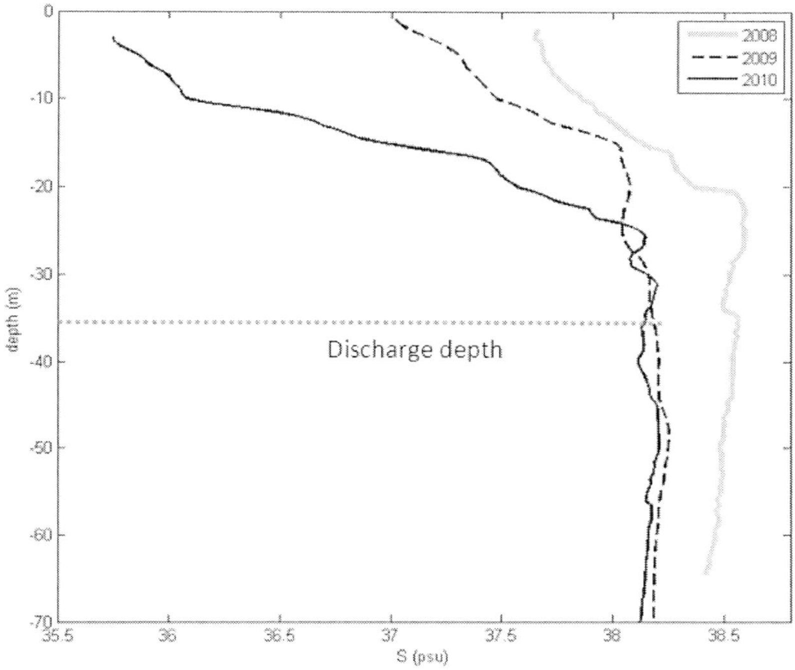

Figure 3. Temperature (°C) (a) and salinity (PSU) (b) profiles measured during the monitoring surveys of 2008, 2009 and 2010.

During the monitoring survey observations of current velocity close to the platform have shown the following surface mean values for the summer period:

2008: $v = 12 \pm 2$ cm s^{-1};
2009: $v = 8 \pm 1$ cm s^{-1};
2010: $v = 10 \pm 1$ cm s^{-1}

these values are compatible with the range (3 – 18 cm s^{-1}) of surface current speed measured in the Adriatic sea by means of Lagrangian drifters (e.g. Falco et al, 2000).

The temperature and salinity profiles measured during the monitoring surveys (Figure 3) indicated strong density stratification every year, typical of summer conditions. The temperature differences over the water column showed little variations among the three surveys and always ranged between ~ 26°C at the surface and 12°C at the bottom, with a thermocline located above the discharge depth. The salinity values showed a maximum between 38 and 38.5 psu for all three years, while a minimum value of ~ 35.8 psu was measured only during the 2010 survey when the water column presented a marked fresh layer located between the surface and ~ 28 m depth (Figures 3 and 4).

The output model results are summarized in Tables 2 and 3 and Figures 5 and 6.

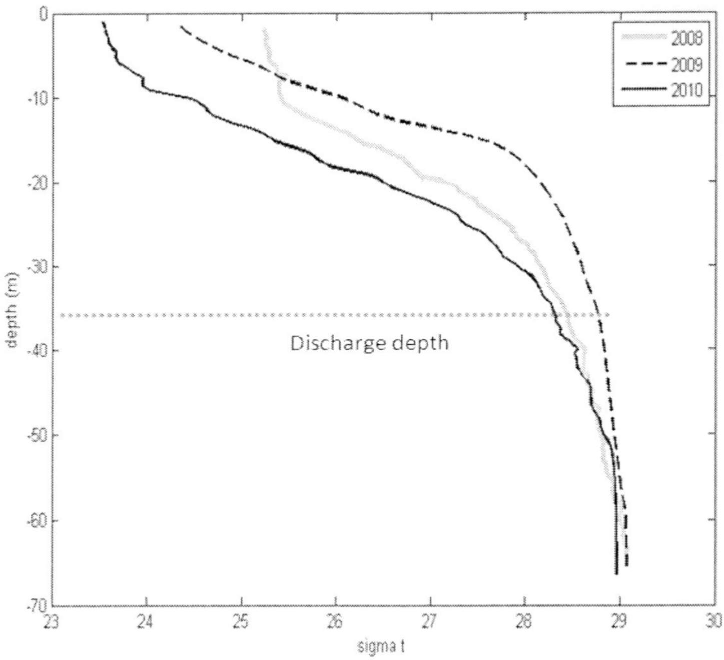

Figure 4. Density profiles in proximity of the selected platform measured during the monitoring survey 2008, 2009 and 2010.

With the (theoretical) diffuser diameter of 45 cm, the effluent showed a positive buoyancy in all three years, thus immediately after the discharge it consistently started to rise within the water column. In the case of 2008 and 2009 field data, the plume reached the trap level in a layer located between 31 and 36 m, while using the 2010 field data the maximum plume rise in the water column was 24 m, where the equilibrium density condition was reached (see Table 2 *a*, *b*, *c*). These rising plume trajectories found in all the simulated scenarios are mainly due to the lower density of the effluent (ρ_{effl} = 26.79 σ_t) than the receiving water at the discharge depth (Figure 4).

Table 2a. Output of near – field simulations for the diffuser diameter of 45 cm and summer 2008 measured field data

Effluent flow mc/s	Trap level depth (m)	Plume diameter (m)	Froude number	Initial dilution Time (h)
fv_1				
2007: 0.0033	35.6	171	0.27	0.04
2008: 0.0029	35.6	165	0.24	0.04
2009: 0.0020	35.6	144	0.17	0.04
2010: 0.0012	35.7	124	0.1	0.04
fv_2				
2007: 0.0038	35.6	179	0.31	0.04
2008: 0.0033	35.6	173	0.28	0.04
2009: 0.0023	35.6	153	0.19	0.04
2010: 0.0014	35.6	129	0.12	0.04
fv_3				
2007: 0.0011	35.7	119	0.09	0.04
2008: 0.0010	33.0	114	0.08	0.04
2009: 0.0007	35.7	100	0.06	0.04

Table 2b. Output of near – field simulations for the diffuser diameter of 45 cm and summer 2009 measured field data

Effluent flow mc/s	Trap level depth (m)	Plume diameter (m)	Froude number	Initial dilution Time (h)
fv_1				
2007: 0.0033	33.8	296	0.25	0.05
2008: 0.0029	33.9	296	0.22	0.04
2009: 0.0020	34.0	262	0.15	0.05
2010: 0.0012	34.3	223	0.09	0.05
fv_2				
2007: 0.0038	33.7	324	0.28	0.05
2008: 0.0033	33.7	309	0.25	0.05
2009: 0.0023	31.9	275	0.2	0.05
2010: 0.0014	34.3	232	0.11	0.05
fv_3				
2007: 0.0011	34.5	231	0.08	0.05
2008: 0.0010	34.5	206	0.07	0.05
2009: 0.0007	34.7	181	0.05	0.05

Table 2c. Output of near – field simulations for the diffuser diameter of 45 cm and summer 2010 measured field data

Effluent flow mc/s	Trap level depth (m)	Plume diameter (m)	Froude number	Initial dilution Time (h)
fv_1				
2007: 0.0033	33.4	339	0.13	0.05
2008: 0.0029	33.6	319	0.11	0.04
2009: 0.0020	24.2	168	0.08	0.7
2010: 0.0012	27.0	132	0.05	0.6
fv_2				
2007: 0.0038	33.2	360	0.15	0.05
2008: 0.0033	33.4	341	0.13	0.05
2009: 0.0023	32.2	493	0.09	0.1
2010: 0.0014	26.4	140	0.05	0.6
fv_3				
2007: 0.0011	27.4	125	0.04	0.56
2008: 0.0010	27.8	118	0.04	0.55
2009: 0.0007	28.8	101	0.03	0.54

Table 3a. Output of near – field simulations for the diffuser diameter of 72 cm and summer 2008 measured field data

Effluent flow mc/s	Trap level depth (m)	Plume diameter (m)	Froude number	Initial dilution Time (h)
fv_1				
2007: 0.0033	35.6	171	0.08	0.04
2008: 0.0029	35.6	164	0.07	0.04
2009: 0.0020	35.6	145	0.05	0.04
2010: 0.0012	35.7	128	0.03	0.04
fv_2				
2007: 0.0038	35.5	179	0.1	0.04
2008: 0.0033	35.6	172	0.08	0.04
2009: 0.0023	35.6	152	0.06	0.04
2010: 0.0014	35.7	135	0.04	0.04
fv_3				
2007: 0.0011	35.7	119	0.03	0.04
2008: 0.0010	35.8	114	0.02	0.02
2009: 0.0007	35.3	101	0.02	0.03

The water column stability leads in all the cases to a densimetric Froude number less than 1, showing a buoyancy-dominated process. The highest values of Fr correspond to the maximum volumes discharged (years 2007 and

2008) for the effluent flow conditions fv_1 and fv_2 (Table 2 *a, b, c*). Using the 2008 and 2009 hydrographic and current data the maximum plume diameter was found in correspondence of the maximum effluent flow and volume (i.e. fv_2 for 2007). Applying the fv_3 condition associated with the very low volume discharged during the year 2010, we found, for all the simulated scenarios, a negligible effect of the PFW on the receiving ambient fluid. For this reason in the following tables the results of the 2010 - fv_3 condition were not reported.

Table 3b. Output of near – field simulations for the diffuser diameter of 72 cm and summer 2009 measured field data

Effluent flow mc/s	Trap level depth (m)	Plume diameter (m)	Froude number	Initial dilution Time (h)
fv_1				
2007: 0.0033	33.7	310	0.08	0.05
2008: 0.0029	33.8	297	0.07	0.05
2009: 0.0020	34	264	0.05	0.05
2010: 0.0012	34.3	224	0.03	0.05
fv_2				
2007: 0.0038	33.6	323	0.09	0.05
2008: 0.0033	33.7	311	0.08	0.05
2009: 0.0023	33.9	274	0.05	0.05
2010: 0.0014	34.2	233	0.03	0.05
fv_3				
2007: 0.0011	34.4	215	0.03	0.05
2008: 0.0010	34.5	207	0.03	0.05
2009: 0.0007	34.6	183	0.01	0.05

A more realistic scenario was simulated using a diffuser diameter of 72 cm; also in this case the summer stability leads to a positive buoyancy plume rising in the water column in all the simulations (Table 3 *a, b, c*). When we used the 2008 and 2009 monitoring data the plume rise height (which corresponded to the trap level depth) was ~ 29 m and 24 m respectively, while the lower density of the surface layer found during the 2010 monitoring survey (Figure 4) induces the plume to reach the trap level at 20 m depth.

Also for the diffuser diameter of 72 cm, for all the monitored years the highest values of the Froude number corresponded to the fv_2 effluent flow condition and to the maximum volume discharged during 2007 and 2008 (Table 3 *a, b, c)*. The summer stratification determines in all the cases Fr values < 1 indicating a plume dominated by buoyancy. The trend of the plume diameter is similar to the Froude number one (Tables 3 *a, b, c,*); independent of the field data used, the maximum values are associated to the highest volume discharged (2007 and 2008).

Table 3c. Output of near – field simulations for the diffuser diameter of 72 cm and summer 2010 measured field data

Effluent flow mc/s	Trap level depth (m)	Plume diameter (m)	Froude number	Initial dilution Time (h)
fv_1				
2007: 0.0033	20.7	216	0.04	0.8
2008: 0.0029	21.7	202	0.03	0.75
2009: 0.0020	24.3	167	0.02	0.68
2010: 0.0012	26.9	133	0.01	0.56
fv_2				
2007: 0.0038	19.9	228	0.04	0.8
2008: 0.0033	20.6	217	0.04	0.8
2009: 0.0023	23.3	178	0.02	0.72
2010: 0.0014	26.4	141	0.02	0.6
fv_3				
2007: 0.0011	27.4	125	0.01	0.55
2008: 0.0010	27.8	119	0.02	0.55
2009: 0.0007	28.8	101	0.01	0.54

Figures 5 and 6 summarize the values of the length of the near field region and average dilution using the field data measured during the monitoring surveys of 2008 (panel a and b), 2009 (panel c and d) and 2010 (panel e and f) for both the diffuser diameter values (45 cm and 72 cm).

The horizontal distance from the discharge point at which the plume reaches the trap level condition (L_d hereafter), represents the length of the near field or initial mixing region. The comparison between the different monitoring years and diffuser diameters showed a high variability of L_d.

When we carried out the simulations with the 2008 field data, independently of the effluent flow conditions, the plume exceeds the limit of acute mixing reaching a horizontal distance of almost 200 m from the source point. For this monitoring year L_d values lower than 200 m are only associated to the diffuser diameter of 72 cm and minimum effluent flow conditions fv_3. Moreover in all the 2008 simulations the plume trajectory did not leave the approximate discharge depth between 35 and 36 m (also see trap level depths in Table 2 *a*).

With the 2009 ambient data the model output showed a very different situation, the scale of the near field region resulting ~15 m for both diffuser diameter values and for all the effluent flow conditions. This suggests that the plume trajectory was mainly forced by the current intensity, which for the year 2009 displayed a mean value lower than the 2008 and 2010 in situ data. Moreover, due to the water column stability, the effluent remained approximately at the discharge depth in a layer between 33 and 36 m depth (Table 2 *b*).

With the 2010 hydrographic and current conditions we found a high variability in the plume horizontal length scales. The maximum L_d ranged from 13.3 to 220 m, with a mean depth of the plume between 24 e 34 m (also see trap level depth in Table 2 *c*).

Comparing the two diffuser diameter conditions it is interesting to notice that adopting the realistic value of 72 cm the plume exceeded the limit of acute mixing zone independently of the effluent flow scenarios. In the case of 45 cm diffuser diameter horizontal distances greater than 100 m are reached with the fv_1 and fv_3 conditions for the volume discharged during the year 2009, with fv_2 only for the 2010 volume, and with fv_3 condition for the years from 2007 to 2009. The maximum width of the initial field region was 257 m and was associated to the maximum volume discharged (2007) for both the fv_1 and fv_2 conditions.

Just like the initial mixing scale also the average dilution factor $S_{a,}$ at the end of the near field mixing process, presented high variable values in the simulated scenarios. All the simulations carried out with 2008 and 2009 field data showed that the initial dilution phase ended within approximately 3 minutes after the release. Using the data of 2008 the initial dilution always reached high values, with a maximum S_a = 727 corresponding to the fv_3 effluent flow condition linked to the maximum volume discharged during 2009. The comparison between the simulations with the two different diffuser diameters showed that in both cases S_a increased at low effluent flow conditions with the exception of the volume discharged by applying fv_3 during the year 2010 and with a diffuser diameter of 45 cm and during 2008-2010 with a diffuser diameter of 72 cm.

The 2009 ambient conditions determined values of S_a (~ 340) are lower than the in the other years; this is mainly due to the effect of the strong density stratification recorded during the summer 2009 (see also Figure 4). The trend of S_a is similar to the 2008 one, with the near field dilution enhanced corresponding to the lowest discharge flows with the only exception being the volume discharged during the year 2010 applying the fv_3 scenario.

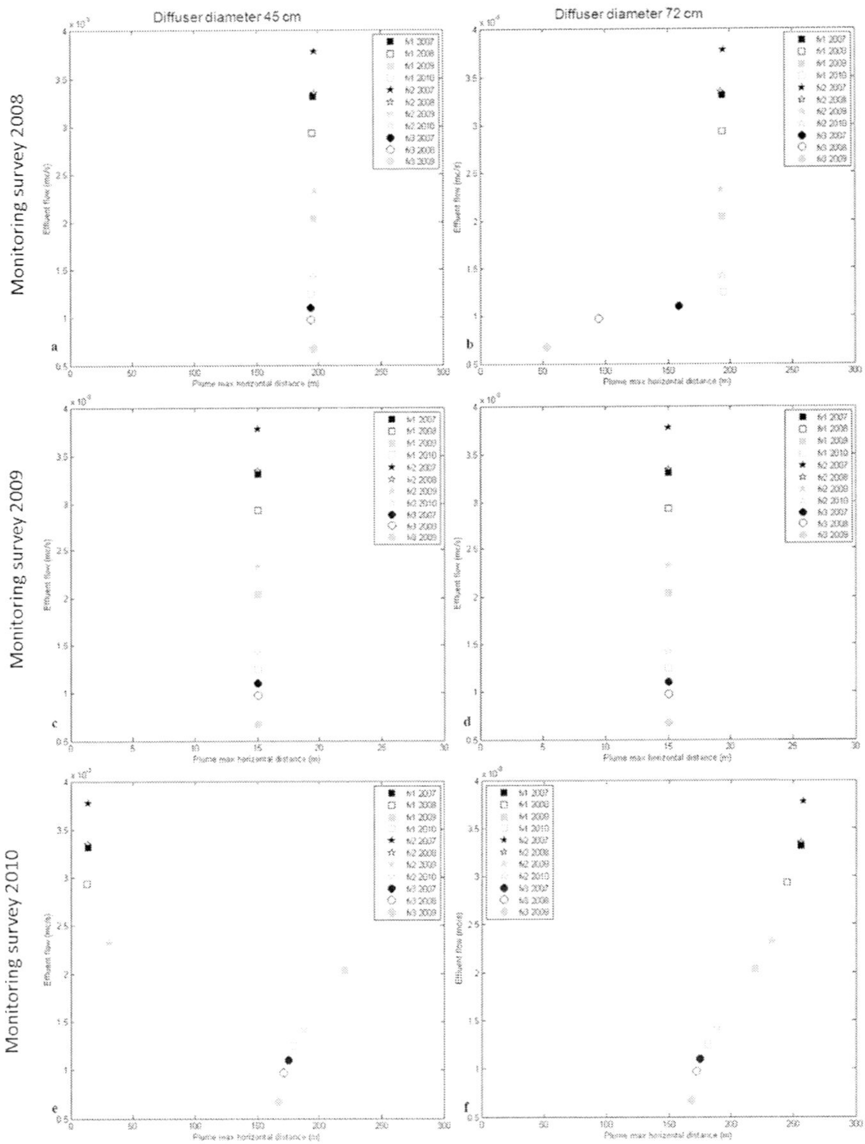

Figure 5. Horizontal distance from the discharge point at which the plume reached the trap level using the field data measured during the monitoring surveys of 2008 (a and b), 2009 (c and d) and 2010 (e and f) for both diffuser diameters (45 cm and 72 cm). The different symbols correspond to the different effluent flow conditions, while different colors correspond to different discharged volumes during the year.

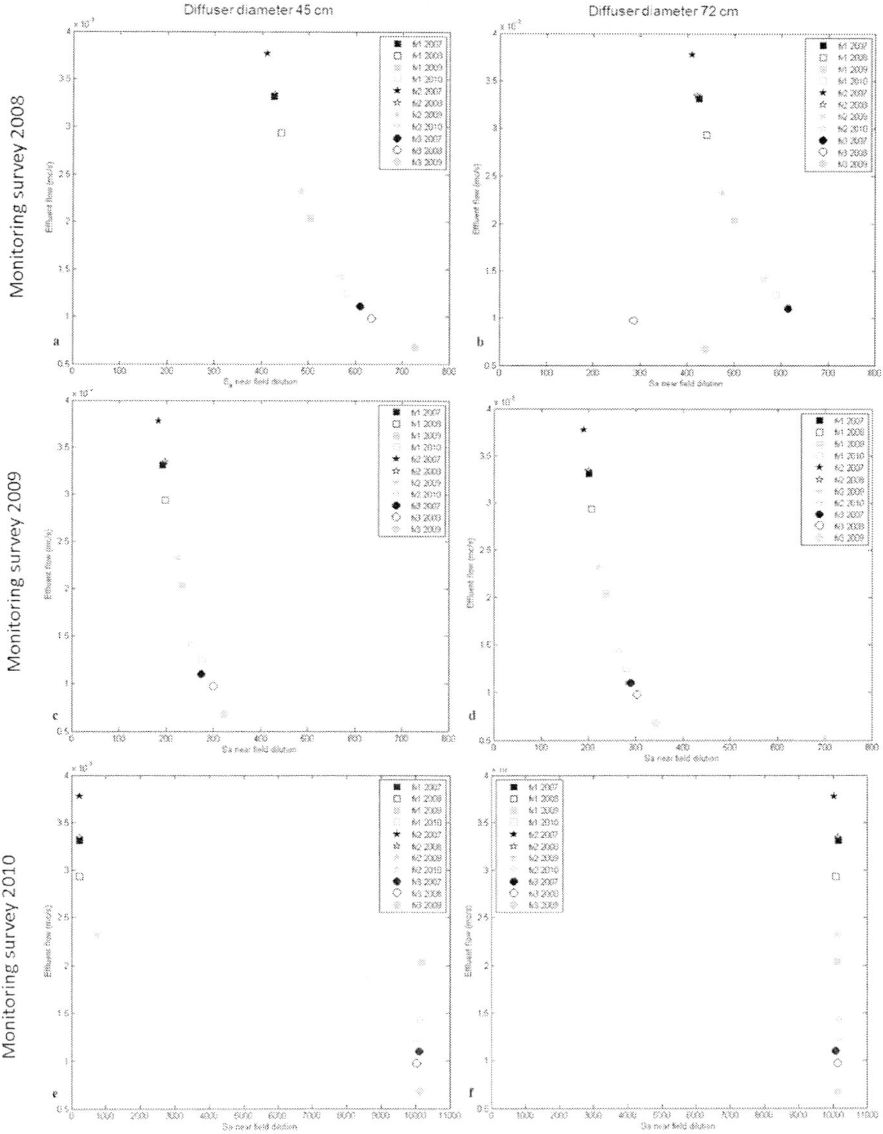

Figure 6. Average dilution factor at the end of the near field mixing process using the field data measured during the monitoring surveys of 2008 (a and b), 2009 (c and d) and 2010 (e and f) for both diffuser diameters (45 cm and 72 cm). The different symbols correspond to the different effluent flow conditions, while different colors correspond to different discharged volumes during the year.

The model results obtained with the 2010 ambient data showed that the average dilution has a behavior similar to L_d for the same monitored year. Comparing it with the 2008 and 2009 model results shows that in this case S_a reached the highest values, thus suggesting a strong forcing effect by both the intense current and the lower density stratification. Adopting the theoretical diffuser diameter, the maximum value of S_a (2280) was reached using the fv_2 condition linked to the maximum volume discharged during the year 2007. In realistic diffuser diameter conditions the average dilution is always > 10000, independent of the discharge flows. It is worth noticing that these high values of S_a always correspond to initial dilution times of ~ 40 min; under these ambient conditions the plume also exceeds the limit of acute mixing zone in ~ 20 min.

CONCLUSION

In the last decades the ever-growing energy demands have been fulfilled by exploiting new natural reservoirs, including offshore oil and gas deposits located in Italian marine coastal areas. This has led to an ever-increasing potential risk for the marine environment surrounding extraction platforms, unless best practices for the extraction operations, strict impact assessment, and careful monitoring procedures are enforced. This is particularly true for areas such as the Mediterranean Sea, which for morphology, dynamics, and socio-economic situations may be subject to multiple anthropic impacts that make them critically sensitive regions from an environmental perspective. The necessity of preserving and sustainably exploiting the environmental resources thus requires accurate studies of pollutant dispersion processes, by which forestalling or mitigating potentially hazardous events.

Here we summarize the results of a work focused on the Adriatic Sea, a sub-basin of the Mediterranean Sea, where a number of offshore natural gas extraction platforms operate. This work was conceived as a parametric study, which applied a multidisciplinary approach (observational-numerical) to investigate the initial mixing of PFWs discharged from selected offshore natural gas platforms. In particular our study dealt with the near-field dispersion processes which take place very close to the outflow, and that are affected by the entrainment processes, and by advection and diffusion in the vicinity of the platform.

Applying a Lagrangian numerical model (UM3) for submarine outfalls we simulated the initial mixing of PFW discharged during the summer season under different hydrodynamic conditions (water column stratification and current intensity) and discharge features (geometrical diffuser characteristic and effluent flow). The present work, integrating observational data from three-year monitoring surveys with different discharge conditions, opens a number of broader perspectives in terms of risk assessment, mitigation, best operational practices and decision policies.

According to previous studies on the Adriatic Sea (Cianelli et al, 2008a; Cianelli et al, 2008b) our results showed that during the monitored summer season the PFW dispersion close to the selected platform was mainly modulated by the seasonal stratification, moreover the PFW plume was mostly spread within a limited layer 3 to 6 meters thick, centered at the source depth. The Froude number values indicated that the water column stability at the platform locations influenced the PFWs dilution more than the local current field, causing in all the simulated scenarios a positive buoyancy plume. These results confirm (see also Cianelli et al, 2008a; Cianelli et al, 2009) that a good option could be represented by releasing higher volumes of PFW under low water column stratification and high mean velocity currents.

The output simulation we obtained also demonstrated that for the same hydrodynamic conditions the different hypothesized effluent flows mainly influenced the PFW dilution factor while only minor effects were reported on the plume's horizontal distance from the source. The dilution factor increased when the low PFW volume released during 2009 was discharged applying the fv_3 effluent flow condition.

On the other hand, the parameter L_d was affected by the diffuser diameter value, in particular under high current intensity and lower water column stability (summer 2008 and 2010).

As far as operational discharge practices are concerned, the results of this parametric study suggest that an optimal release scenario might be to continuously unload the PFWs three times a week in the space of 24h each time.

This discharge practice should promote the rapid dilution of the effluent, mitigating the impacts of the PFW on marine environment. In summary, all the previously described results at present may play a crucial role in a "prevention first" policy and represent an important first step in the design of a decision-making action to protect marine ecosystems from the impacts of offshore platform operations.

REFERENCES

Baumgartner, D. J., Frick, W. E., Roberts, P. J. W.; *Dilution models for effluent discharges (3rd Edition)*, Pacific Ecosystems Branch, ERL-N, Newport, 1994; EPA/600/R-94/086.

Berry, J. A. In *Offshore Oil and Gas Environmental Effects Monitoring - Approaches and Technologies*; S. L. Armsworthy, P. J. Cranford, & K. Lee (Eds.), Battelle Press: Columbus, 2005; 111-129.

BP; *What's inside? BP Statistical Review of World Energy;* 2011; pp.49- bp.com/ statisticalreview

Cianelli, D., Manfra, L., Zambianchi, E., Maggi, C., Cappiello, A., Famiglini, G., Mannozzi, M., Cicero, A. M.; *Mar. Env. Res.*, 2008a; 65, 325-337.

Cianelli, D., Scotti, T., Zambianchi, E., Manfra, L., Cicero, A.M.; *XVII AIOL Conference Proceedings*, 2008b; 19(1), 137-141.

Cianelli, D., Manfra, L., Maggi, C., Zambianchi, E.; *Exploration and Production: Oil and Gas Review* 2009; 7(2), 129-132.

Cianelli, D., Manfra, L., Zambianchi, E., Maggi, C., Cicero, A.M. In: *Fluid Waste Disposal*, Kay W. Canton (Ed.), 2011; Chapter 4, 113-135, Nova Science Publishers, Hauppauge, NY.

Cushman-Roisin, B., Gacic M., Poulain, P-M., and Artegiani, A. *Physical Oceanography of the Adriatic Sea. Past, Present and Future*; Kluwer Academic Publishers, Dordecht / Boston / London 2001.

Di Mento, R., Mannozzi, M., Mafra, L., Gulia, P., *Linee guida per la redazione del piano di monitoraggio volto a verificare "l'assenza di pericoli per le acque e per gli ecosistemi acquatici" derivanti dallo scarico diretto a mare delle acque risultanti dall'estrazione di idrocarburi*, 2009 ISPRA, www.isprambiente.it.

Enerdata, *Global Energy Statistical Yearbook* 2011- yearbook.enerdata.net.

Falco, P., Griffa, A., Poulain, P., Zambianchi, E.; *J Phys. Ocean.*, 2000; 30, 2055–2071.

Frick, W.E.; *Env. Model. Soft.*, 2004; 19, 645–654.

Frick, W.E., Roberts, P.J.W., Davis, L.R., Keyes, J., Baumgartner, D.J., George, K.P.; *Dilution models for effluent discharges, 4th ed.*, Visual Plumes. Ecosystems Research Div., NERL, U.S. EPA, 2002; EPA/ 600/R-03/025, Athens, GA.

Manfra L., De Nicola, E., Maggi, C., Zambianchi, E., Caramiello, D., Toscano, A., Cianelli, D., Cicero, A.M.; *J. Mar. Biol. Assoc. UK* 2010; doi:10.1017/S0025315410001037.

Maggi, C., Trabucco, B., Mannozzi, M., Manfra, L., Gabellini, M., Di Mento, R., Nonnis, O., Virno Lamberti, C. and Cicero, A.M.; *Rapport de la Commission Internationale pour l'Exploration Scientifique de la Mer Méditerranée*, 2007; 38, 688.

Manfra, L. and Maggi, C. In *Advanced in Natural Gas Technology*, H.A. Al-Megren (Ed), InTech, Croatia, 2012; 271-284.

Manfra, L., Moltedo, G., Virno Lamberti, C., Maggi, C., Finoia, M.G., Gabellini, M., Giuliani, S., Onorati, F., Di Mento, R. and Cicero, A.M.; *Arch. Environ. Contam. Toxic.*, 2007; 53, 183-190.

McCormack, P., Jones, P., Hetheridge, M.J., Rowland, S.J.; *Water Res.*, 2001; 35 (15), 3567–3578.

OGP *Fate and effects of naturally occurring substances in produced water on the marine environment*, International Association of Oil & Gas Producers, 364/2005.

OSPAR *Assessment of impacts of offshore oil and gas activities in the North-East Atlantic*; Ospar Commission, London, 453/2009, ISBN 978- 1-906840-93-8, 40 pp.

Pinder, D.; *Ocean & Coastal Manag.*, 2001; 44, 579-600.

Roberts, P. J. W., Snyder, W. H., and Baumgartner, D. J.; *J. Hydr. Engr.*,1989; ASCE,115(1), 1-25.

Roberts, P.J.W., Tian, X., *Env. Model. Soft.*, 2004; 19, 691–699.

Russo, A., and A. Artegiani, *Sci. Mar.*, 1996, 60(2), 33-43.

Scott, K.A., Yeats, P., Wohlgeschaffen, G., Dalziel, J., Niven, S. & Lee K.; *Mar. Env. Res.*, 2007; 63, 146-167.

Trabucco B., Maggi, C., Manfra, L. Nonnis O., Di Mento R., Mannozzi M., Virno Lamberti C., Cicero A.M. and Gabellini M. In *Advances in Natural Gas Technology*, H.A. Al-Megren (Ed), InTech, Croatia, 2012; 285-300.

Wilson III, C.A. & Heath, J.W. In *Encyclopedia of Ocean Science*, Steele J.H., Thorpe S.A., Turekian K.K.(Eds), Academic Press, Orlando, 2008; 2414-2419.

In: Mediterranean Sea
Editor: Terrence B. Hughes
ISBN: 978-1-62618-238-7

Chapter 4

SECCHI DEPTH AND WATER QUALITY CONTROL: MEASUREMENT OF SUNLIGHT EXTINCTION

Javier Ibáñez Civera[1,2]***, Nicolás Laguarda Miró***[1,3,*]***,***
Eduardo García Breijo[1,2] ***and Rafael Masot Peris***[1,2]
[1]Instituto Reconocimiento Molecular y Desarrollo Tecnológico. Universitat Politècnica València, Valencia, Spain
[2]Departamento de Ingeniería Electrónica. Universitat Politècnica València, Valencia, Spain
[3]Departamento de Ingeniería Química y Nuclear. Universitat Politècnica València, Valencia, Spain

ABSTRACT

Water is fundamental for life and its quality control is becoming more and more important. In fact, there are vast strategic actions and laws trying to manage it. An example is the European Water Framework Directive 2000/60 EC that summarizes and manages all the actions to water quality control in European Water bodies. In this sense, Turbidity is one of the water parameters that may present some problems when being measured. This is due to the fact that the Secchi Disk measure is a worldwide accepted procedure but it may have some limitations such as

[*] Corresponding Author's Email: nilami@iqn.upv.es.

the subjectivity of the measure (depends on the ability of the operator), its dependence on environmental factors and its limitation in shallow water bodies where it may reach the bottom without having disappeared.

The goal of this chapter is to collaborate in the improvement of turbidity measurement by suggesting an alternative procedure to Secchi Depth determination based on a new device called LUZEX. This new device uses optoelectronic measurements of sunlight extinction with depth in water bodies to calculate a sunlight extinction coefficient for different wavelength bands.

Several measurements have been done in continental, sea and transitional water bodies in order to carry out a comparison between the measured values for Secchi Disk and the obtained values for Secchi depth based on a mathematical model and the data given by our LUZEX device.

As a conclusion, the suggested method to substitute Secchi Disk measurement is quite good as we have obtained regression coefficients (R^2) up to 0.9967 in our best models.

1. Introduction

Light is the essential physical factor for phytoplankton growth, as photosynthesis depends entirely on solar energy, which is regulated in the aquatic environment by the the sunlight penetration in water. Light intensity decreases with depth, and after 100 meters it is insufficient for phytoplankton to perform the photosynthesis.

Sunlight penetration in seawater has allowed scientists to divide the ocean into three vertical zones: euphotic, aphotic and disphotic zones:

The euphotic zone is well illuminated, wherein the amount of light entering the water is sufficient to perform photosynthesis. In the open ocean this zone contains water from the surface up to 100 m depth. In the mouth of the rivers and estuaries, where river waters still flow, light can only penetrate a few meters, as the waters that bring these rivers are loaded with suspended matter such as dust, sand, phytoplankton, etc., that hinder the penetration of the light rays.

Moreover, the water is agitated and troubled by the action of waves, currents and tides. However, almost the entire phytoplankton community is located in this area due to the abundance of nutrients.

The Disphotic area, poorly illuminated, is located in the sea between 100 and 200 m depth and the amount of light is weak and insufficient to have a net production of organic matter. This area only has some phytoplankton when it is carried by the water currents.

The aphotic zone has no light. Darkness reigns and heterotrophic organisms dominate this region. This is the largest area of the oceans and it starts at 200 meters and reaches great depths up to 11,000 meters.

Natural waters are complex (Wozniak and Derak, 2005). The incident light on the surface undergoes changes in its propagation direction. The amount of light that passes through depends on the surface status, the sunlight intensity and the water properties. Also the surge and the nature of the surface itself (film surface) decrease the effective radiation within the water column.

The light available for photosynthesis in water bodies is affected by two processes: absorption and scattering (Kirk, 1996; Mobley, 1996). Absorption completely eliminates light from water while dispersion changes its direction of propagation. Scattering does not eliminate light from water but it increases the probability of being absorbed by the increase of the distance that light must run through the water. Absorption and scattering interact in a complex and nonlinear way (Wen-Cheng, 2005). Equations governing the light propagation underwater are called radiative transport equations and they have no exact solution but there are several computer programs that have been designed to solve these equations by different numerical methods (Mobley et al. 1993). Despite the complexity of the radiative transport equations, the irradiance measurements underwater usually shows a negative exponential attenuation of light with depth.

In the absence of strong discontinuities in water quality as nephelometric layers, submerged chlorophyll "a" layers or suspended organic matter, the measurement of photosynthetic active radiation (400-700 nm) is well described by a single exponential equation (Lambert-Beer equation, Dennison et al. 1993)[2]:

$$Iz_2 = Iz_1 \cdot e^{-Kd \cdot (Z_2 - Z_1)}, \tag{1}$$

where Iz_1 and I_{Z2} are irradiances at z1 and z2 depths respectively. Kd is the extinction coefficient:

$$Kd = -\frac{1}{z} \cdot \ln\left(\frac{I_Z}{I_0}\right) \tag{2}$$

The extinction coefficient of light in water is due to both absorption and scattering.

Then, the light received by a water body comes from the sunlight direct radiation and the radiation scattered by suspended particles in the water body (diffuse radiation).

Overall, Kd is the sum of the absorption coefficients (a) and dispersion (b), all of them depend on the wavelength of light (λ):

$$Kd(\lambda) = a(\lambda) + b(\lambda)\,, \tag{3}$$

2.Sunlight Extinction in Water

When light propagates in an aquatic environment, its intensity decreases exponentially due to attenuation phenomenon which has the previously cited main causes: absorption and scattering.

These two phenomena are explained next:

Light Absorption

The light energy is converted into another type of energy, usually heat or chemical energy. This absorption is caused by:

- Phytoplankton.
- Organic and inorganic suspended particulate matter.
- Dissolved inorganic compounds.
- Water.

These substances cause a series of disturbances in the sunlight rays as they can be absorbed in some of the incident radiation wavelengths and they can also change their trajectory.

Absorption Due to Phytoplankton

Phytoplankton consists of a series of aquatic autotrophic organisms which have the ability to carry out the photosynthesis. There also exist thousands of phytoplankton species that can be classified according to several factors: color, size, structure, ecological niche, etc.

These phytoplankton species (algae and some bacteria) are characterized by having chlorophyll, so photosynthesis can be carried out. In this process, energy is captured as light to be transformed into chemical energy and thus grow as living matter.

Photosynthesis takes place in two stages: in the first one (primary phase or light) a series of reactions that depend on the light and are independent of temperature are carried out, while the second (secondary or dark phase) reactions depending on temperature but independent of light are carried out.

The speed of the first stage increases with light intensity (up to certain limits), but not with temperature. In the second stage, the speed is increased with temperature but not with the light intensity.

Light plays an important role in the first of these two stages and this is due to chlorophyll (organic compound formed by molecules that contain carbon, hydrogen, oxygen, nitrogen and magnesium atoms) that captures sunlight and causes the breakdown of water molecules separating the hydrogen from the oxygen. In other words, the chemical bond that holds these elements in the molecule of water, breaks down because of the light.

Therefore, phytoplankton is the first step in the pelagic world's food chain and this is the reason why it has such a high ecological value.

Absorption Due to Dissolved Organic Matter

Dissolved organic matter, phytoplankton pigments and inorganic matter are other important compounds that may affect to the light absorption in water and they influence in the overall absorption spectrum of light in the range of the visible radiation.

Dissolved organic matter in water bodies is mainly formed by humus compounds (yellow substance) and its absorption spectrum decreases exponentially with the wavelength of the radiation. In addition, it is also composed by other substances (proteins, carboxylic acids, carbohydrates, vitamins, amino acids, peptides, phenolic compounds, lipids, etc.), that are also responsible for the attenuation of light in the water column in the wave range corresponding to the ultraviolet radiation.

Moreover, the amount of light absorbed and the spectrum wide range in which the absorption is maximum, also depend on the concentration and composition of the dissolved organic matter.

Absorption Due to Dissolved Inorganic Compounds

On the other hand, the presence of dissolved inorganic compounds must be considered as another important factor that affects the behavior of light in the water column.

These compounds affect the transparency of the water (Babin et al, 2003) and usually produce a dispersion of light that change the trajectory of increasing the same way that the light should go, favoring the absorption by other components present in the water.

The extinction of light in the water due to these substances depends also on other factors (as with the dissolved substances) such as its composition, thickness and shape. Therefore the greater inorganic compounds greater the light extinction in the water.

Absortion Due to Water

Water is an essential substance for the development of all the vital activities of living organisms. It is a molecule consisting of the establishment of the "hydrogen bonds" between them and other substances, due to the formation of dipoles by electrostatic Van der Waals forces. These "hydrogen bonds" are formed due to the partial negative charge of the oxygen that carries electrostatic attraction on the partial positive charges of hydrogen atoms. Although unions are weak, the fact that there are four other molecules around every molecule of water via the "hydrogen bonds" allows the water to have a lattice-like structure.

Therefore, the water molecule will also take part in the sunlight absorption even though it is lower than the absorption due to phytoplankton, organic matter and inorganic compounds. Then, as water will affect the light extinction, it must be considered when calculating the total light extinction coefficient, K.

Light Scattering

This phenomenon is due to the collision of a light beam with suspended particles in water (inorganic matter from the outside or sediments), causing multiple reflections. The more turbid the water is (more particulates), the greater scattering phenomena there will be. The problem raises in the fact that

Scattering prevents the light penetration in water affecting to the normal development of different species of plants and animals that live in these environments. In addition to the concentration of the particles, Scattering also depends on other factors such as their composition, size and shape.

Other Factors

The light extinction in water (absorption and scattering) is often caused by other factors. Bioluminescence, fluorescence and bubbles are three clear examples of this kind of factors that also affects the presence of light in the water column.

The presence of these other factors may be casual so their presence may produce errors when measuring if they are not taken into consideration. Then, it is advisable to consider them when calculating the sunlight extinction coefficient. Below, a brief description of these three fenomena is given:

Bioluminescence

Bioluminescence is a phenomenon by which certain organisms (bacteria, fungi, insects, fish, etc.) produce light. In some species, it is used to communicate, to attract preys, to aid in reproduction or as a defense to confuse predators.

However, this feature, which sometimes is a major consumer of energy for these organisms, can be seen in the depths and in the dark, but is insignificant if there is sunlight.

Fluorescence

This is the phenomenon that happens if a substance generates light when exposed to ultraviolet radiation, cathode rays or X-rays. The absorbed radiation (invisible to the human eye) is transformed into visible light.

In this process, a molecule absorbs a high-energy photon, which is emitted as a photon of lower energy (higher wavelength). The energy difference between absorption and emission is dispersed as heat (molecular vibrations).

When this phenomenon happens in water, it's usually due to the presence of chlorophyll and organic matter (yellow substance).

Bubbles

This phenomenon occurs due to changes in the surface of a water body when there is movement that make air bubbles of different sizes enter inside the water. Bubble generation and properties depend on the magnitude of the movement. The presence of bubbles in the water column may cause a light backscatter higher than the water itself.

3. Secchi Disk and Sunlight Extinction

Secchi Disk is a device created by Pietro Angelo Secchi in 1865. It is an easy and economic device used to measure transparency in water bodies such as lakes, oceans, rivers or bays, and it is widely used all around the Mediterranean.

On the other hand, the conditions to carry out the Secchi Disk measurements must be established in order to facilitate future data comparison. A good procedure is to make measurements on calm days between 9 am and 3 pm. In addition, measures should be made from the leeward side of the boat with the sun at the back of the person that is measuring.

Although it is considered a good method for the Mediterranean and its coastal, transitional and continental water bodies, Secchi Disk measurements do not provide an accurate measure of transparency since there may be errors due to the subjectivity of the human eye, due to different environmental conditions during the day, the inclination of the sun, the state of the water, time of day, weather, etc., so it is recommended to have measure all the samples by the same person. This is, in fact one of the disadvantages of the Secchi Disk that can be improved by using optical devices.

4. LUZEX

The Group of Electronic Development and Printed Sensors. (GEDPS) of the Institute of Molecular Recognition and Technological Development (IDM) at the Polytechnic University of Valencia has designed and developed a new device to measure the sunlight extinction in water bodies (Ibáñez et al, 2011 a).

The device called "LUZEX" consists of two different units:

1. Measurement probe.
2. Monitoring and storage equipment.

Figure 1. Submersible measuring probe..

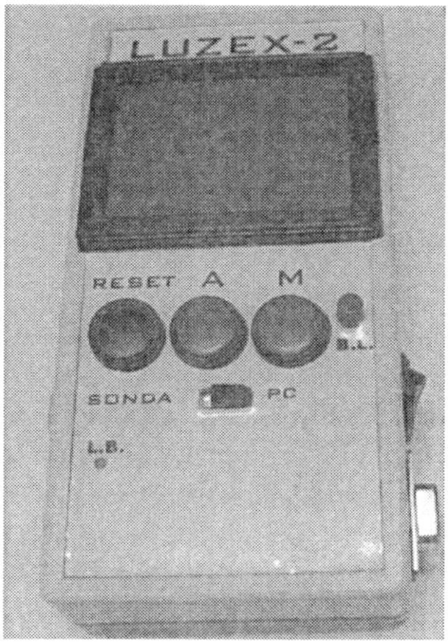

Figure 2. Monitoring and storage equipment.

The main features of the equipment are:

- Measurement of sunlight intensity in water at 365 (UVA), 460 (B), 540 (G), 660 (R) and 900 (IR) nm.
- Measurement of temperature between 2 and 150 ° C.
- Measurement of pressure from 0 to 30 psi (approx. 2 atm).
- IP68 submersible (10m. depth for 2 weeks).

- Monitoring equipment and storage: 10h autonomy and capacity for 8 different sampling points with approx. 500 samples / point.
- PC Communication via RS232 serial port USB convertible.

4.1. Measurement Probe

The probe contains the photodiodes sensors and the signal adaptive systems (amplifiers and filters). A microcontroller is responsible for the measurements and send the data via RS-485 communication to the monitoring and storage equipment.

In addition to the photosensitive elements, the probe includes a temperature gauge (integrated) and a pressure sensor. Both measures are also sent to the monitoring and storage equipment. The system is enclosed in a sealed PVC with a transparent glass window to allow light to enter. In the end, all fasteners are made of stainless steel to prevent oxidation in aggressive media.

4.2. Monitoring Equipment and Storage

The monitoring device comprises a display screen as a user interface. In addition, it also contains the program menu, the navigation buttons and the power supply system, comprising a rechargeable battery, charging terminal and the external supply connection. Other devices are also present such as a RS232 output for PC connection and a switch-on/off button.

The system has been constructed in a commercial polyethylene case. The equipment is controlled by a Microchip 16F877 microcontroller. In addition, the system also has peripherals such as a RS-485 receiver (to receive the information from the probe), a RS232 serial port for communication with the PC, a graphic display and EEPROM memory system mass storage for the acquired data.

The software allows two operation modes:

- Automatic. The measurements are stored at intervals of 0.5 seconds.
- Manual. The measurements are saved with a press of the user. The monitor show the values for light intensity, pressure and temperature, and the user decides which sample data must be saved or rejected.

In order to perform measurements at different sampling points, the information is saved in a different memory banks each time that a new measurement starts, providing enough memory to store data from up to 8 sampling points. Each one of the sampling points can store approximately 500 measurements.As LUZEX carries out measures to calculate the sunlight extinction coefficients in water for the five studied wavelengths, a mathematical expression for Secchi Depth can be obtained based on these coefficients. Then, attending to the state of the art in these correlations, some authors (Jamu and Piedrahita, 1999) use the following formula:

$$K_d = \frac{K_e}{SD}, \tag{4}$$

where Ke is a constant whose value is between 1.1 and 2.0 and SD is the Secchi depth.

In order to estimate Secchi depth using different light coefficients extinction by LUZEX, the mathematical expression will be as follows:

$$SD' = \frac{\alpha}{B1 \cdot K(UV) + B2 \cdot K(R) + B3 \cdot K(G) + B4 \cdot K(B) + B5 \cdot K(IR)}, \tag{5}$$

where SD' is the estimated Secchi depth. K (UV, R, G, B, IR) are the different sunlight extinction coefficients for each used wavelength, and finally, α and B1-B5 are the different coefficients experimentally determined which will be obtained by statistical methods.

5. Three Different Mediterranean Aquatic Ecosystems

In order to obtain representative results, three different water areas were considered. That is, saltwater, freshwater and wetland water bodies. We found it interesting to analyze the differences in these waters because of their own nature.

As known, the sea water contains dissolved inorganic salts, mainly sodium chloride. Approximately, the ocean water salinity is 35000 ppm (35 g /l). While fresh water salinity is notably lower: less than 1000 ppm (1 g /l).

Moreover, it is expected that both wetlands and freshwater areas are more polluted than the seawater due to turbidity (created by currents, erosion, vegetation, land ...). It was decided to design a sampling area near the city of

Valencia as an example of Mediterranean waterbodies and we decided the sampling points looking in each case for the most symbolic or representative water to be sampled. So was taken L'Albufera as an example of Mediterranean wetland. L'Albufera Natural Park is located 10 km south-east from the city of Valencia. Beaches of Port Saplaya and Pobla Farnals, located at the northeast of the city, were choosen as representative coastal sea waters. Finally, we considered the Turia River, northwest of Valencia., as an example of continental freshwater zone. A brief description of these areas is shown next:

L'Albufera

L'Albufera Natural Park is one of the most valuable and important wetlands around Valencia and in the Mediterranean in general. It currently has an area of 21,120 hectares, comprising the municipalities of Valencia, Massanassa, Sedaví, Alfafar Albal, Catarroja, Silla, Sollana, Beniparrell, Sueca, Cullera, Algemesí and Albalat de la Ribera.

It is located 10 km south-east from the city of Valencia and it includes the system formed by the lagoon itself, its surroundings and the humid coastal strip adjacent to the Mediterranean shore.

It was declared a Natural Park by the Regional Government on July 23th, 1986 and is protected by several laws and declarations belonging to the lists of ZEPA ecological areas and the RAMSAR list of the European Union.

Historically, the area has been populated by humans, being at present surrounded by rice fields and a series of population centers and industrial parks that have been created in recent decades and have generated an enormous pressure on the water environment. In fact, this waters are characterized by their eutrophic aspect and the specific algal blooms in spring and summer due to the significant nutrient content in the water. This phenomena causes anoxia and fish kills as phytoplankton blooms occurs. Meanwhile, its waters present a very low transparency.

The River Turia

The Turia is a typical Mediterranean river which rises in the Iberian Range and flows in Valencia city having a length of 280 km. This river, like all of the area flow differs greatly at different seasons, with an average flow of 14 m3/s in Vilamarxant but having reached peaks above 3.000m3/s in episodes of large floods that inundated Valencia. The water inflows occur primarily by streams and ravines caused by the thaw in the birth and severe storms that occur in the watershed especially during episodes of a typical autumn climatic phenomena known as “Gota fría” (Cold drop) characterized by deep depresions and strong located rainfalls in the region.

Overall, its waters have a good quality and are highly transparent, so that area has been chosen for sampling as it is representative for the typical Mediterranean inland water bodies.

Beaches of the Valencian Coast

Valencian coasts, especially those closer to the capital city, are open and wide fine sand beaches entering the sea with a smooth profile and they have quiet and pristine waters. This is due to the properties of the Mediterranean in this region, particularly the absence of major river inputs and soft marine currents that run from north to south that make their water bodies to be homogeneous, transparent and generally of a very good quality.

The significant number of Blue Flags that the European authorities give to Valencian beaches year after year as well as the many tourists who come in summer to enjoy these beaches, are also indicatives of their quality and recognition.

Moreover, analyses and studies periodically conducted in these waters show very good physical-chemical quality, a major algal biodiversity and a rich diversity of marine life especially in areas where the *Posidonia Oceanica* is the basis for their livelihood.

Therefore, we have chosen some Valencian coastal water bodies as the ideal representative of Mediterranean beach.

6. Procedures

6.1. Measured Parameters

Secchi Depth

Light extinction with depth was measured in the determined sampling points with the designed equipment. Likewise, Secchi Disk was used for determining water transparency in centimeters as measured with this technique.

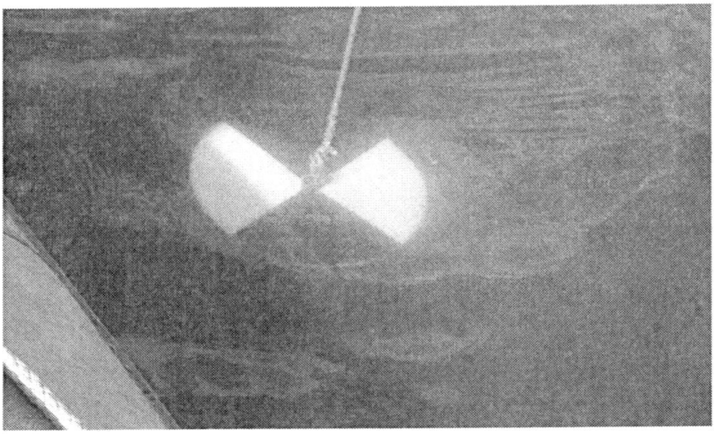

Figure 3. Secchi Depth measurement in a sampling point at L'Albufera.

The procedure for measuring this parameter is the following: First, let the disk submerge vertically in the water until its division lines cannot be seen. This depth is written down as the Descending Sechhi Depth. Next, pull up the rope until the Secchi Disc divison lines can be seen again. Then write down this depth as Ascendingh Secchi Depth. In the end, the value for Secchi Depth in a determined sampling point is the average of these two measurements made.

Measurement with Luzex

The procedure consisted on the installation of the complete LUZEX equipment, and the immersion of the sensor slowly in the water controlling its descent with a rope. When the LUZEX is ready over the water surface, the monitor is swithced on and then we can start collecting measures of sunlight extinction in the monitoring and storing device while the measurement probe goes down until reaching the bottom of the water body. After reaching the bottom at each sampling point, we must repeat the same procedure but while the measurement probe is ascending to the surface. Then, all the data is collected and LUZEX can be switched off to go to the next sampling data to continue collecting data (Ibáñez, 2009).

Later, we can download the collected data into a computer to perform the necessary calculations and statistical data treatment.

6.3. Data Treatment

Data acquired with LUZEX are sent through the serial port to a computer in order to work with them in an EXCEL file. The Solver application from EXCEL was used to obtain coefficients and carry out the adjustments to the equations of the sunlight extinction in each studied wavelength. The regression coefficient R^2 has been taken into consideration to assure an appropriate adjustment of the model to the measured values for Secchi Depth by using the Disk.

In order to determine the different extinction coefficients data are transmitted via the serial port to an Excel file with a format like the one shown below:

Table I. Example of data downloaded from LUZEX to an Excel file

Section	LPRES	LTEMP	LUV	LR	LG	LB	LIR
0	570	522	285	245	365	285	379
0	572	522	250	130	321	180	243
0	574	522	95	141	235	132	156
0	575	522	54	135	194	104	116
0	577	522	29	123	162	83	88
0	578	522	17	110	139	65	67
0	578	522	9	100	123	55	57
0	579	521	7	93	113	48	49
0	580	522	5	87	104	42	43
0	580	522	3	82	96	36	37
0	581	522	2	77	88	32	33
0	582	522	1	72	81	27	29
0	583	522	1	67	75	25	26
0	582	522	1	67	75	25	25
0	584	522	0	61	67	20	21

6.3.1. Extinction Coefficients

The calculation of these different extinction coefficients for each of the wavelengths, is based on the calculation of partial least squares for the mathematical expression shown next. It allows us to calculate light intensity for each of the wavelengths and compare it with the data given by LUZEX

$$I_z = I_0 \cdot e^{-K_c \cdot x} \quad (6)$$

where: Iz and I0: light intensities for depths z and 0 respectively.

Kc: extinction coefficient for each of the bands of different wavelengths

x: depth.

Furthermore, the different Kc are obtained by using the Excel Solver. The calculation starts using an arbitrary value of Kc, and next, a least squares fit between the sampled values of each illumination of the photodiodes and the values obtained with the previous formula, will finally let us know the Kc values.

6.3.2. Secchi Depth Calculation

As seen before, to determine the extinction coefficient from the measurements of Secchi depth at each sampled point, we use the following equation:

$$Kd = \frac{Ke}{SD}, \quad (7)$$

where: Kd: extinction coefficient

Ke: constant with a value between 1.1 and 2 (as the author)

SD: Secchi Depth measured directly with the Secchi Disk

The idea of this section is to calculate the Secchi depth estimation by measuring the light extinction, in order to verify if it is possible to replace the Secchi Disk for LUZEX

$$SD' = \frac{\alpha}{B1 \cdot K(UV) + B2 \cdot K(R) + B3 \cdot K(G) + B4 \cdot K(B) + B5 \cdot K(IR)}, \quad (8)$$

The estimation of the parameters α and ßi, also called Secchi depth coefficients is done experimentally using the necessary adjustments with the Excel Solver applying the above mentioned formula and the obtained extinction coefficients for each of the studied wavelengths (Ibáñez et al, 2011 a).

Next, the data obtained with the Secchi disc for each of the sampled points along Valencian region are shown:

Table 2. Examples of sampling points, Secchi Depth measured and additonal data

Location	Point	Section (LUZEX)	Latitude (N)	Longitude (O)	Date	Hour	SD down (cm)	SD up (cm)	SD average (cm)
Canal de Club Náutico de Pinedo	2	2	39°25'38"	0°20'11"	25/09/2012	17:00	35	65	50
Club Náutico de Pinedo	3	4	39°25'41"	0°19'57"	25/09/2012	17:15	160	155	157,5
Azud en el río Tuéjar	4	5			26/09/2012	14:10	45	45	45
Carraixet 1	5	0	39°30'08"	0°19'31"	29/09/2012	13:00	77,5	77,5	77,5
Carraixet 2	6	3	39°30'07"	0°19'37"	29/09/2012	15:00	100	100	100
Port Saplaya 1	7	5	39°30'37"	0°19'16"	29/09/2012	16:00	150	150	150
Puerto deportivo de Pobla de Farnals	8	6	39°33'40"	0°17'04"	01/10/2012	16:45	170	170	170
Puerto deportivo de Canet de Berenger	9	7	39°40'26"	0°12'10"	01/10/2012	17:15	150	150	150
Gola de Pujol 1	10	1	39°20'58"	0°19'00"	01/10/2012	15:00	20	20	20
Gola de Perellonet	11	3	39°18'35"	0°17'48"	01/10/2012	15:40	20	20	20
Embalse de Buseo	12	0	39°35'12"	0°56'44"	03/10/2012	15:10	80	80	80
Embalse de Buseo	13	1	39°35'13"	0°56'47"	03/10/2012	15:20	90	90	90
Carraixet 3	14	1	39°30'07"	0°19'35"	16/10/2012	13:40	77,5	77,5	77,5
Puerto deportivo Port Saplaya 2	15	2	39°20'34"	0°19'06"	16/10/2012	14:10	180	180	180
Gola de Pujol 2	16	3	39°20'57"	0°19'06"	16/10/2012	15:00	30	25	27,5
Canal El Palmar 1	17	5	39°19'01"	0°19'02"	16/10/2012	15:35	20	20	20
Canal El Palmar 2	18	6	39°18'59"	0°19'73"	16/10/2012	15:45	25	25	25

As shown in the graphic below, the infrared (IR) wavelength presents a sharper slope. This fact means that it disappears faster than the rest.

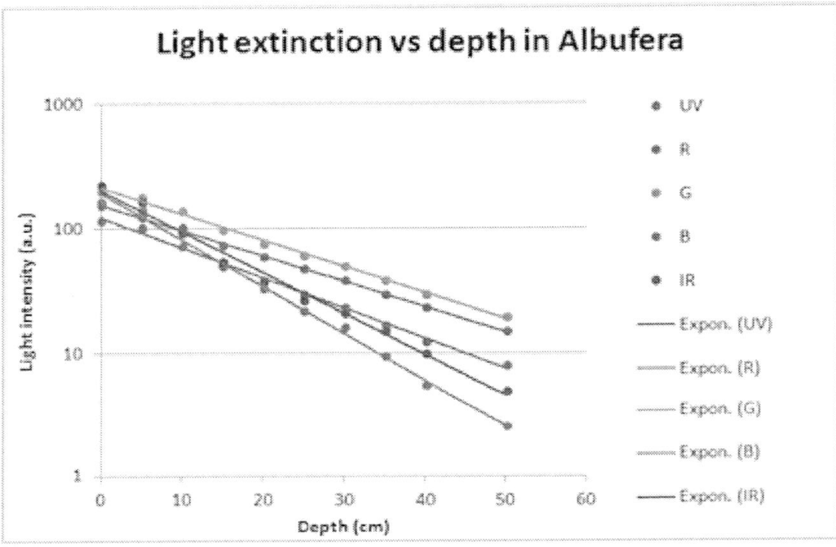

Figure 4. Light extinction versus deep in Albufera lagoon. Straight lines shows fitted exponentials.

Table 3. Extinction coefficients and R2 coefficient

Sensor	λ (nm peak)	Adjusted exponential	R^2	Extinction coefficient (cm^{-1})
Ultraviolet (UV)	375	$y = 122,14e^{-0,056x}$	0,9966	K(UV)= 0,056
Red (R)	660	$y = 154,06e^{-0,047x}$	0,9988	K(R)= 0,047
Green (G)	540	$y = 212,49e^{-0,049x}$	0,9959	K(G)= 0,049
Blue (B)	460	$y =192,85e^{-0,087x}$	0,9952	K(B)= 0,087
Infrared (IR)	850	$y =201,69e^{-0,076x}$	0,9903	K(IR)= 0,076

For each sampled point, graphics like the one shown are obtained. Extinction coefficients for each of the wavelengths studied are obtained with the data and using Solver. In this way, a table is shown with the obtained extinction coefficients for each of the bands at each sampling point, and the measurement of Secchi Depth obtained using the Secchi Disk:

Table 4. Secchi depth at some sampling points and extinction coefficients (obtained by Solver) for each of the studied wavelengths

Location	Point	K_{UV}	K_{RED}	K_{GREEN}	K_{BLUE}	K_{IR}	SD measured (cm)
Canal de Club Náutico de Pinedo	2	0,0509	0,0188	0,0136	0,0205	0,0699	50
Club Náutico de Pinedo	3	0,0110	0,0103	0,0083	0,0121	0,0401	157,5
Azud en el río Tuéjar	4	0,0873	0,0375	0,0478	0,0532	0,0483	45
Carraixet	5	0,0265	0,0145	0,0153	0,0181	0,046	77,5
Carraixet	6	0,0873	0,0375	0,0478	0,0532	0,0483	100
Port Saplaya	7	0,0103	0,0226	0,0125	0,0188	0,0636	150
Puerto deportivo de Pobla de Farnals	8	0,0081	0,0102	0,0084	0,0109	0,0508	170
Puerto deportivo de Canet de Berenger	9	0,0157	0,0122	0,0093	0,0131	0,0751	150
Gola de Pujol	10	0,6632	0,0441	0,0552	0,1221	0,0661	20
Gola de Perellonet	11	0	0,0220	0,0360	0,0371	0,0551	20
Embalse de Buseo	12	0,0116	0,0098	0,0086	0,0101	0,0294	80
Embalse de Buseo	13	0,0534	0,0105	0,0077	0,0111	0,0378	90
Colector Carraixet	14	0,0462	0,0121	0,0052	0,0150	0,0655	77,5
Puerto deportivo Port Saplaya	15	0,0269	0,0111	0,0125	0,0133	0,0496	180
Gola de Pujol	16	0,0802	0,0164	0,0112	0,0177	0,0272	27,5
Canal del Palmar	17	0,0802	0,0705	0,0783	0,1058	0,0931	20
El Palmar	18	0,1704	0,0723	0,0722	0,1327	0,0723	25

In addition, LUZEX lets us to detect discontinuities in the water column. It is specially interesting in the river mouths where continental waters meet the saltwater and a stratification is generated that is not possible to detect with the traditional Secchi Disk method. In this way, the use of LUZEX gives us the opportunitie to monitor the river plumes at their mouth, knowing the depth of the discontinuities in the water body properties. An example of this kind of monitoring is shown in Figure 5 for a typical mixture of continental and marine waters in the Valencian coast (Almenara channel, Valencia).

If we make a least squares adjustment of the Secchi Disk depth measured with the Secchi Depth estimated by using data from the LUZEX device

(Ibáñez et al, 2011 b), we obtain the following graph showing that the correlation coefficient is 0.9227. In addition, if other more complex correlation methods are used, the value for R2 can be improved. This is the case we have developed by Artificial Neural Networks. In this specific analysis, a net of neurons has been designed to calculate the values for the Secchi Disk Depth based on the sunlight extinction coefficients given by LUXEZ for each studied wavelenght. The designed artificial neural network is able to develop a self-analysis and learn for itself in order to modify the mathematical model and adapt it to find the best value for R2. Finally, in our Artificial Neural Network-based calculation for Secchi Depth, a value of 0.9967 is obtained for the Correlation Coefficient.

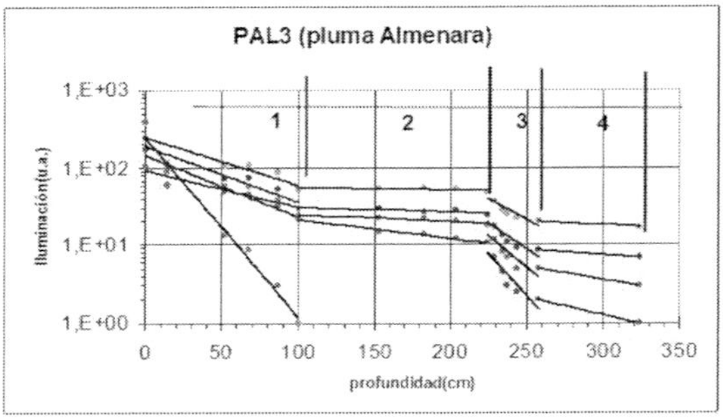

Figure 5. Water stratification and discontinuities in sunlight extinction at the Almenara Plume (Valencia).

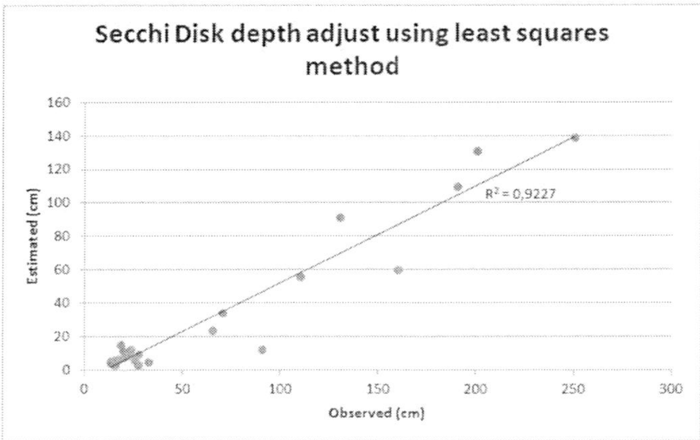

Figure 6. Correlation among observed (Secchi Disk) and Estimated (LUZEX) values by means of Leasts Squares.

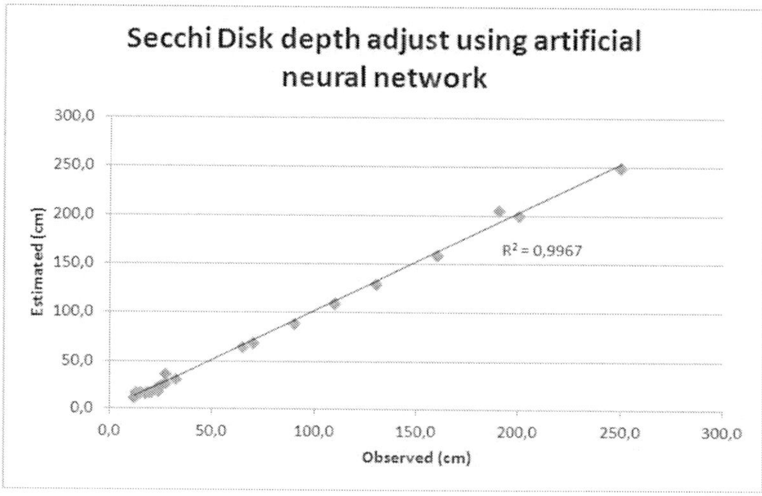

Figure 7. Correlation among observed (Secchi Disk) and Estimated (LUZEX) values by means of Artificial Neural Networks.

Finally, it is interesting to highlight two major advantages that this method has over the traditional Secchi Disk. First, it eliminates the subjectivity of the measure carried out directly by the observer, so that the error due to the human eye disappears. Second, LUZEX lets us determine water clarity in terms of Secchi Disk Depth in shallow water bodies where the disk reaches the bottom before it disappears from the view of the observer, a fact that makes it impossible to determine Secchi Depth by the traditional method . In these cases, LUZEX allows performing the measurements. Thereby, the measure of the transparency of a water body by calculating the coefficients of extinction of light in water is considered an advance in the state of the art of monitoring this water quality parameter and we are convinced that its use will be implemented in the future.

CONCLUSION

This study developed in Valencian water bodies has determined the light extinction coefficients in the water by using the LUZEX device. These light extinction coefficients for each of the five bands of the spectrum (UV , BLUE, RED, GREEN, IR) show a clear correlation with the measured Secchi Depth in all sampled points.

It has been observed that there is a significant presence of suspended solids in most of the water sampled. This fact is demonstrated by the values obtained for the extinction coefficient in the wavelength range corresponding to the infrared. And a low presence of organic matter, showed by the low values of the extinction coefficients for the photodiodes of ultraviolet and blue wavelengths with respect to the infrared.

Furthermore, it was possible to calculate the estimated Secchi depth, SD', by using the extinction coefficients. In addition, this has allowed to calculate Secchi Disk depths from the sunlight extinction coefficients obtained by LUZEX. One can say that the adjustment is very good, since the best value of R2 obtained has a value of 0.9967. We want to highlight that the measure carried out "in situ" by LUZEX, let's us to detect changes in water composition along the column and it is very useful in monitoring the quality of water bodies suffering stalking discharges from different continental sources. This has been seen in some of the graphs showing the light intensity with depth. In these graphics, it was observed a change of the slope of the straight line at a certain depth, which is due to a stratification of the water column and therefore, it means that there is a variation in water composition along the column. In this case, stratification of the water column has happened in the locations of the Puerto deportivo de Pinedo, in the Barranco de Carraixet and Port Saplaya. In the end, it has been found that it is possible to substitute the Secchi Disk for the LUZEX device to measure water transparency. This not only would avoid possible errors arising from the use of a method as subjective as the Secchi Disc is, but it would also to obtain a broader and more diverse range of parameters that would allow a more thorough and complete study of these kinds of mediterranean water bodies.

Acknowledgments

Authors acknowledge financial support from the Spanish Government project MAT2009-14564-C04-02.

References

Babin et al. 2003. Variations in the light absorption coefficients of phytoplankton,nonalgal particles, and dissolved organic matter in coastal waters around Europe. *Journal of Geophysical Research,* VOL. 108, NO. C7, 3211, doi:10.1029/2001JC000882, 2003.

Dennison WC, Orth RJ, Moore KA, Stevenson JC, Carter V, Kollar S, Bergstorm PW, Batiuk RA (1993). Assessing water quality with submersed aquatic vegetation: habitat requirements as barometers of Chesapeake Bay health. *Bioscience,* 43:86–94.

Ibáñez Civera, J. Desarrollo de un sistema de medida de la extinción de la luz solar en medios acuáticos. Caracterización de la composición y distribución de las particulas suspendidas a lo largo de la columna de agua. TESIS DOCTORAL. Universitat Politècnica de València. 2009.

Ibáñez Civera J., Romero Gil I., Laguarda-Miro N., Garcia-Breijo E., Gil-Sánchez L., Martínez-Guijarro R.. Instrument for sunlight extinction measurement in water bodies. *Sens. Actuators A,* 168 (2011) 267–274.

Ibáñez Civera J., Garcia Breijo E., Laguarda Miró N., Gil Sánchez L., Garrigues Baixauli J., Romero Gil I., Masot Peris R., Alcañiz Fillol M.. Artificial neural network onto eight bit microcontroller for Secchi depth calculation. *Sens. Actuators B,* 156 (2011) 132–139.

Jamu DM, Lu Z, Piedrahita RH (1999) Relationship between Secchi disk visibility and chlorophyll a in aquaculture ponds. *Aquaculture,* 170:205–214.

Kirk, J.T. 1996. Light and photosynthesis in aquatic ecosystems. Cambridge Univ. Press, Bristol. 509 pp.

Mobley, C. D., (1996). Light and water, Academic Press, Ch 1-4.

Mobley CD, Gentili B, Gordon HR, Jin Z, Kattawar GW, Morel A, Reinersman P, Stamnes K, Stavn RH (1993) comparison of numerical models for computing underwater light fields. *Appl. Opt.,* 32:7484–7504

Wen-Cheng Liu. Water column light attenuation estimation to simulate phytoplankton population in tidal estuary. *Environ Geol.,* (2005) 49: 280–292 DOI 10.1007/s00254-005-0087-y.

Wozniak B. and Deraq J. (2007). Light absorption in sea water. *Springer-Science.*

In: Mediterranean Sea
Editor: Terrence B. Hughes

ISBN: 978-1-62618-238-7

Chapter 5

AN ENVIRONMENTAL MONITORING PLAN FOR THE CONSTRUCTION AND OPERATION OF A MARINE TERMINAL FOR REGASIFYING LIQUEFIED NATURAL GAS (LNG) IN THE NORTH ADRIATIC SEA

C. Virno Lamberti, M. Gabellini, C. Maggi, O. Nonnis, L. Manfra*, S. Ceracchi, B. Trabucco, G. Moltedo, F. Onorati, G. Franceschini and R. Di Mento

ISPRA Institute for Environmental Protection and Research
Rome, Italy

ABSTRACT

Natural gas is mainly used for generating electricity and for residential uses. Italy does not extract a sufficient amount of natural gas to meet domestic consumption. Given the growing demand for natural gas and the need to differentiate supply routes to make Italy less dependent on individual countries, in 1998 the project of the "Adriatic LNG Terminal" was authorized. The project consists of the construction and operation of an offshore terminal for regasifying Liquefied Natural Gas (LNG) and its pipeline for transferring it to land, in the Northern

* Corresponding author: loredana.manfra@isprambiente.it.

Adriatic Sea (Porto Viro, Rovigo). The Italian Ministry of the Environment has assessed the environmental compatibility of this project prescribing monitoring to verify possible impacts on the marine environment associated with the construction and operation of the structure. The Institute for Environmental Protection and Research (ISPRA, formerly ICRAM) elaborated and carried out the monitoring plan according to environmental prescriptions. The monitoring involves environmental studies in the marine area (where the terminal and part of the pipeline is located) and in the wetland area consisting of a lagoon and a fish valley, crossed by the connecting pipeline before reaching the land. Three monitoring phases have been put on: 1) previous to the construction of the structures, 2) during the construction and 3) during the terminal's operation.

This paper describes the Adriatic LNG Terminal and its pipeline Monitoring Plan (ALNGM). This plan provides the application of an integrated approach in order to analyze the possible environmental alterations that could occur.

INTRODUCTION

Natural gas, produced by the anaerobic decomposition of organic material and retrieved from underground, is used primarily for the production of electricity and for residential use. Italy does not extract a sufficient amount of natural gas to meet domestic demand. Until a few years ago the main importing countries were Russia, the North Sea, Libya and Algeria. The increasing demand for gas and the need to differentiate supply routes to make Italy less dependent on individual countries, are the main strategic reasons for the development of offshore LNG regasification Terminal project located in the Northern Adriatic Sea.

The project, started more than 10 years ago and realized by Adriatic LNG (company established by Edison, Qatar Petroleum and ExxonMobil), consists of the construction and operation of an offshore terminal and the connection's pipeline to the Italian gas network. The Terminal regasification operation, started in 2009, covers about 10% of the current natural gas domestic requirements and creates a new way for gas imports from the Persian Gulf, the place of production [1]. This is the first offshore LNG terminal in Italy and the first in the world Gravity Based Structure (GBS) for unloading, storing and regasifying.

The terminal, located approximately 12 km away from the nearest coast, consists of a reinforced concrete structure (GBS) resting on the seabed at a depth of about 29 meters. The GBS 180 mt. long, 88 mt. wide and 47 mt. high, was built in Spain and was towed by tugboats to its current location [2]. Once it reached its place of location, the sinking of the structure was carried out by injection of a quantity of seawater necessary to rest it on the bottom. Subsequent to this phase of "provisional ballasting", the "definitive ballasting" occurred, through the use of sand mixed with sea water and pumped into the compartments of GBS, with sedimentation of solids contained in it. The ballast water, before final discharge into the sea, was filtered to reduce the suspended solids. Once the GBS was placed in the final position, in order to protect the structure from erosion, a belt composed of large size blocks was laid on the seabed, along the entire GBS perimeter. The submerged belt has been designed to ensure maximum settlement of flora and fauna on hard substrate according to the Italian authorization requirements [3, 4]. Moreover, in order to promote the repopulation of commercial marine species in the Terminal area in accordance with the requirements [3, 5], ALNGM designed and realized artificial reefs at about 1.5 km from the Terminal in Southwest direction. The LNG (Natural Gas cooled to -162°C permitting a reduction in the volume of 600%), transported from the production site by special tanker ships, is downloaded to the terminal where it is stored, regasified, and then transported to the ground, through the pipeline, to be entered into the distribution network [6]. The pipeline consists of an off-shore section, along approximately 15 km (from the terminal to the coast), and a on-shore section, along approximately 25 km. Sealine's installation provides for laying on the seabed and covering. The terminal's regasification plant is located on top of the GBS. It consists of vaporizers that operate using the natural heat of water withdrawn from the sea during the regasification process. The seawater used for regasification is discharged (maximum flow rate of 29000 m^3/hr) in the South direction at a depth of -12 m with a temperature lower than the receiving system (thermal delta of -4,6°C). To prevent the growth of vegetation in the circuits, the withdrawn seawater is treated with sodium hypochlorite (for antifouling); the active chlorine concentration at the discharge (0,2 mg/l) is in accordance with Italian regulations [3, 7].

In 2005, on behalf of Adriatic LNG, the Institute for Environmental Protection and Research (ISPRA, formerly ICRAM) that acts under the vigilance and policy guidance of the Italian Ministry for the Environment, designed the monitoring plan for the marine environment, identifying the main impacts linked to the project [8].

In this paper we report on the Adriatic LNG Terminal and its pipeline Monitoring (ALNGM). First, we describe the main possible impacts associated with the construction and operation of the regasification terminal, then we detail the matrices and analyses considered and the sampling strategy used.

MAIN POSSIBLE IMPACTS OF TERMINAL AND PIPELINE TRANSFERRING GAS TO THE LAND

Terminal Impacts

During construction and operation phases, the terminal may interfere with the marine environmental quality around the structure.

The terminal's construction could determine potential impacts connected both to the ballasting operations to anchor it to the bottom and to the physical presence of a considerable size structure. The chemical characteristics of sand used for the ballasting have to be guaranteed, but the process may introduce particles with an increase of suspended solids and turbidity in the water column; for this reason, the presence of a great offshore structure on the seabed certainly leads to changes in local water flow, in the rate of erosion or sedimentation, modification in bottom morphology, variation in grain size, and organic content in sediments. Additionally, causing disturbance of the seabed induces a consequent sediment resuspension, mobilization of chemical substances (e.g. organic matter, nutrients, metals and pollutants in general) and a temporary increase in turbidity due to the material raised from the bottom [9]. The terminal's construction could affect the marine communities and the responses of organisms living in the area. Impacts to marine biological resources (pelagic and benthic) are considered significant if they cause adverse changes in community composition or ecosystem relationships for important species [10]. However, the terminal can be an element of uniqueness in the morphology of the Northern Adriatic Sea as it may act as a shelter for fish and other mobile marine organisms, and provide a habitat for benthic organisms usually associated with hard substrates. On the basis of this consideration, the terminal is considered not only a subtraction of habitats but also new substrate availability.

In the operation phase the regasification process causes the most significant impacts. The heat exchange waters, discharged into the sea, appear to be cooler and with a chlorine residual that could give rise to organic halogenated derivatives. The difference of temperature can cause disturbances to marine organisms while some chlorinate organic substances can be dangerous for biota [9]. The offshore terminal also determines maritime traffic and noise, particularly in the supply vessels and the tankers.

Pipeline Impacts

The pipeline construction involves the removal of sediment layers to create the trench where the pipeline is laying. The interaction with the seabed results in: a) a change in the morphology; b) a sediment resuspension, resulting in a suspended solid increase in the water column (turbidity increase); c) a recirculation of the absorbed matter, including possible polluting substances (e.g. organic compounds, metals). Turbidity increase can adversely affect the biomass production, particularly for bivalve mollusks (clams) during the breeding season. Contaminant bioaccumulation can occur in aquatic filtering organisms living close to the bottom. However, conditions of minimum mixing of the water column reduce the diffusion of resuspended sediment. The laying of the pipeline on the seabed may alter the structure of biocenosis. Mortality of some benthic species or loss of breeding sites for these species can occur in the area of submerged pipeline. Biocenosis should gradually recover when the pipeline is covered. The pipeline, once in the bottom, is not dangerous for the marine environment, therefore during the terminal operation phase its presence should not lead to significant alterations. However, the progressive deterioration of the pipeline may cause, in time, accumulation of constituent elements of the pipeline that may interfere with sediment quality.

Adriatic LNG Terminal and Its Pipeline Monitoring (ALNGM)

ALNGM was born as an integrated plan. Matching information on physical, chemical and biotic variables, the plan tried to give the best description of the environmental quality status [11, 12, 13] and then to identify spatial-temporal trends of the possible alteration that could occur in relation with the project. The main aims of the program were: a) to monitor in space and time selected chemical and physical variables of the water column and sediment, also by remote sensing analyses; b) to evaluate possible effects on biota (by bioassays and biomarkers, bioaccumulation analyses, community analyses) and living organisms (by bioacoustics investigations); c) to study bottom modifications by acoustic surveys and taking images by remote operated vehicle. The integrated approach hence is to be meant without spatial "fractionizing": in order to make a valid assessment, from an ecological point of view, of the possible effects due to the construction and operation of structures, studying the offshore marine ecosystem together with the inshore one is crucial.

The offshore area is characterized by a homogeneous morphology of the seabed. The continuity of this environment is interrupted by the presence of scattered organogenic carbonate rocks, locally known as "Tegnùe" or "Tresse" (isolated structures, emerging from North Adriatic sediments in a wide area from Grado to the Po river).

These rock formations are environments of particular importance from an ecological point of view, increasing biodiversity of the Adriatic seabed: they allow the establishment of local unique benthic fauna and flora of hard substrate, substantially different from those found in the surrounding soft sediments.

They are, therefore, sensitive situations that may be adversely affected by both natural and anthropogenic phenomena such as resuspension of sediment. Investigation and environmental monitoring of these environments assume, therefore, particular importance in order to assess the possible impacts over the short- or long-term, determined by the construction and operation of the terminal and its pipeline. The wetland environment consists of the Vallona lagoon and Bagliona valley. The lagoons are special environments, at which transition takes place between land and sea, and the mixing of fresh water with the marine water creates a mosaic of different habitats. These habitats are populated by a surprising variety of life forms that are classified among the highest diversity in ecosystems. The Ramsar Convention [14] recognizes that wetlands constitute a great economic, cultural, scientific and recreational resource, recommending the study, management and rational use. For these reasons, the European Normatives (e.g. [11, 15]) target precisely the preservation and conservation of transitional waters. The Vallona lagoon overlooks the Adriatic Sea, in an area between the mouth of the Po di Levante (at the north) and the mouth of the Po Maistra (at the south); the most important activities are the intensive and extensive clam farming. The onshore area is surrounded by a continuous series of fish farms, which draw from the lagoon most of the water necessary for their farming activities, including the

Bagliona valley. The Vallona lagoon and the Bagliona valley, detailed above, are interested in the construction and operation of pipeline.

Monitoring activities are therefore designed to contain the possible impacts and to ensure environmental protection. Three monitoring phases have been provided: 1) previous to construction of the facilities, 2) during the construction and 3) during the terminal's operation. The phase, previous to the construction of the facilities, provides an overview of the environmental state "pristine" compared to human activities that must take place. Thanks to this "temporal blank", it is possible to compare data obtained during this phase with those gained during following phases.

The phases during the construction and the operation, respectively, take a picture of the system in crucial moments, when the unfolding of human activities can cause possible impacts on the studied ecosystem. The Plan provides measurement of several variables in different periods: physical-chemical, chemical, sedimentological, geomorphological, ecotoxicological, ecological, bioacoustics and remote sensing ones.

In order to obtain an exhaustive and rigorous environmental framework, all analyses are also carried out in a "control area", sited in a geographic proximity so as to reflect the same local, natural, physical, chemical and biological conditions of the monitoring site but so far from this site as not to be affected by the activities that happen in it. The monitoring is devised as a flexible tool allowing modifications when the activities are in progress, in relation with the acquired results.

MATRIX SAMPLING

ALNGM involves investigations on different environmental matrices: water, sediments, biota and sea bottom.

Water

Physical, chemical and biological properties are investigated in the water column to study the possible short-term changes. Physical-chemical parameters are monitored in situ by multiparametric CTD probe, equipped with additional specific sensors. Water samples for chemical analyses are collected using a "rosette" device equipped with Niskin bottles, sampling at different depths. Phytoplankton and bacteria are also sampled using Niskin bottles, zooplankton by a plankton net. In addition, satellite observations are used to monitor, by means of passive remote sensors, some physical-chemical parameters such as superficial temperatures, chlorophyll, dissolved organic materials and suspension solids.

Sediment and Sea Bottom

All substances, generally, are accumulated in the seabed and particularly in the sediment. Sediment is a conservative matrix and its investigation is very important to better evaluate the environmental quality status and identify eventual spatial trends of possible alterations. Sediments are collected by Van Veen grab or box corer, at the different sampling stations, treated and stored according to definite protocols (e.g. [16]). Bottom morphology and batimetry characteristics of the project offshore area are investigated using a Side Scan Sonar and a Multibeam instrument, respectively; visual surveys are also performed by a Remotely Operated Vehicle (ROV).

Biota

Contaminants bioaccumulation and some early warning responses (biomarkers) are analyzed in "bioindicators" living in the investigated areas. For example *Mytilus galloprovincialis* is a good bioindicator widely used in monitoring plans. ALNGM allows use of mussels from the farm located near the pipeline and mussels collected from a reference site (Portonovo, AN). They are transplanted for 4-6 weeks in the investigated areas nearby the GBS. In addition native bivalves are collected in the lagoon area. In the offshore area the species of bivalves *Chamalea gallina* and in the lagoon the species *Tapes philippinarum* are used as bioindicators. The fish community in the area around the terminal is also monitored by catch analysis, and the fish species farmed in the Bagliona Valley were subjected to bioaccumulation and biomarker analyses.

Sampling Strategy

The sampling strategy is different for terminal (offshore area) and pipeline (offshore and wetland areas) according to the specific potential impacts to study. Moreover, it changes in function of the matrix and of the monitoring phases (Table 1).

To study contamination and benthic community structure around the terminal area, water and sediment were sampled at different distances from it, with a radial design with a predominant number of sample stations in the main marine current direction.

Table 1. Sampling strategy details for water, sediment and biota

Area	Matrix	Sampling Frequency	Monitoring phase
Terminal	water	six-month/annual	C, O
	sediment	annual	B, C, O
	biota	Seasonal/six-month	B, C, O
	water	annual	C, O
Pipeline	sediment	annual	B, C, O
	biota	Seasonal/six-monthly	B, C, O

B: Blank, C: Construction, O: Operation.

This design permits us to highlight environmental changes when the point source of disturbance is known [17,18]. Mussels are transplanted in a nearby area and in a reference site further away from the terminal. Fish are also collected nearest to the terminal and in a reference site. The fish population study was developed in the whole area around the terminal (macroscale study) and in the proximity of the structure (microscale study). In the pipeline area sediments were sampled along its route.

The sample stations nearest to the coast are at smaller distances than offshore stations; waters are collected only in a part of the sediment sample stations. Specific analyses on the water (e.g. suspended solids), on the hard bottom zoobenthic community and on fish assemblage are provided on "*tegnùe*" close to pipeline to monitor possible alterations caused by the construction phase.

Geomorphological and remote sensing analyses are provided as supplementary investigations, both in the terminal and in pipeline areas during the three phases, while bioacoustics are only close to the terminal in order to monitor eventual noises.

ANALYSES

The analyses have been chosen considering a) the main impacts due to the LNG Terminal Project, b) the characteristics of investigated areas, c) the normatives and authorization requirements related to the project.

Physical-Chemical Analysis on the Water Column

During the construction and operation phase there are specific possible variations of water column physical-chemical characteristics. In fact, the movement of sediments during sealine and GBS positioning may cause an increase of turbidity and suspended solids. Additionally, during the vaporization process the discharge with a thermal delta of -4.6°C may vary the local temperature profile. Furthermore the processes that undergo the effluent discharged into the sea also depend on the water column stratification. For these reasons, the study of physical and chemical characteristics of the water column play an important role in ALNGM. Physical-chemical data of the water column are collected by means of a multiparametric probe [19] with the following sensors: temperature, salinity, pH, dissolved oxygen, redox potential, fluorimeter and turbidimeter. In the marine area around the GBS current data have been acquired by means of the ADCP (Acoustic Doppler Current Profiler) to assess the hydrodynamism variation due to the presence of the structure. Considering the thermal delta associated with the vaporization process and the sodium hypochlorite used in the circuits as antifouling, the fate of cold and chlorinated water's discharge is also studied by means of satellite observations and processing the physical-chemical and current data to implement a numeric model estimating effluent dilution [20].

Chemical Analysis

The purpose of chemical analysis is to assess eventual contaminations caused by construction and operation of the terminal and its pipeline. So, the chemical parameters investigated in water, sediment and biotas have been selected, taking into account the main impacts mentioned above for the offshore structures.

Besides, the chemical parameters have been selected considering some important regulations for the issue of: a) the seabed movement as a result of cables and laying of pipelines [21], b) compliance with quality standards for sediments of coastal waters ([22] and c) protection of the aquatic environment [11, 23].

The monitoring plan provides the chemical analyses reported in Table 2.

Among them, there are some key parameters for the monitoring of these structure impacts, for example: a) suspended solids and organic matter in the water column to evaluate the dispersion of resuspended sediments and b) disinfection byproducts in water, sediment and biota to assess the impact due to the discharge of chlorinated water produced during the regasification process.

Table 2. Chemical analysis details for water, sediment and biota

Water	Sediment	Biota
Nutrients, total hydrocarbons, disinfection byproducts, particulate organic matter, suspended solids, bacteria contamination	PAHs, PCBs, organotin compounds, metals, pesticides, total hydrocarbons disinfection byproducts, organic matter, total organic carbon, bacteria contamination (in the wetland area PCDD and PCDF, estrogenic compounds too)	PAHs, PCBs, organotin compounds, metals, pesticides, disinfection byproducts, bacteria contamination (in the wetland area dioxins, furans and fish estrogenic compounds too)

Disinfection byproducts (such as trihalomethanes, haloacetic acids, halophenols etc.) are priority substances and are included in the list of substances most at risk for the water sector. Bioaccumulation analysis mainly concerns contaminants, which are particularly interesting in areas of intense activity of shellfish, fish farming and of natural bivalve exploitation, above all for marine organisms representing an economic resource. In addition, the chemical speciation of certain substances (such as trace elements) to evaluate their mobility and bioavailability is investigated. It is very important that all chemical investigations are carried out according to standardized procedures and/or scientifically recognized.

Sedimentological Analysis

Sediment grain size allows us to study trends in surface processes related to the dynamic conditions of transportation and deposition and so observe eventual modifications caused by the structures. Moreover, the relationships between physical properties and chemical characterization of sediment are a basic point to determine eventual contaminations related to terminal and pipeline activities.

Geomorphological Analysis

The study of the sea bottom morphological characteristics, by geophysical devices (e.g. Side Scan Sonar and Multibeam echosounder), allows us to highlight the relationship between natural features and the presence of anthropic structures. This analysis represents an essential tool for accurate study of the impact of human activities on the seabed [24, 25]. The acoustic investigations supply the background information necessary for correct planning of sampling activities and represent the basis for a long-term monitoring study. Furthermore, they allow us to monitor the sea bottom geomorphological modifications and also map the sea bottom alterations due to the terminal and its pipeline presence.

The area for geomorphological investigations is performed in a square shape area (9 NM^2) around the terminal and in a rectangular shape area (15Km long and 1Km wide) around the pipeline with a suitable range of acquisition of 100 m.

Ecotoxicological Analyses

A series of biological analysis tools are employed to monitor the environmental effects of construction and operational phases. Some aspects related to these investigations are briefly described.

Biomarkers

Biomarkers are any biochemical, cellular, physiological or behavioral changes in tissues, fluid samples, or at the level of whole organisms, resulting from environmental and/or chemical exposure [26] and are intensively applied in monitoring programs as early warning signals of contaminant exposure and /or of potential ecosystem degradation caused by contaminants [27]. Different biomarker analyses measured in different target species (sentinel organisms) allow detection of different effects. Biomarkers measured in fish and mussels, allow evaluation of effects correlated to contaminants that are in the water column while biomarkers are measured in benthic mollusks (e.g. *Tapes philippinarum* and *Chamelea gallina*) allowing assessment of toxic effects correlated to contaminants absorbed in sediment. A battery of biological responses selected among "biomarker of stress", "biomarker of genotoxicity" and "biomarker of exposure" were analyzed in appropriate tissues (Table 3). Such biological responses can be influenced by natural factors and this basal variability needs to be distinguished by the effects caused by pollutants [28]. Further, statistical analyses (e.g. ANOVA or U-Mann Whitney and PCA) are applied in order to discriminate between different sites and/or samplings.

Table 3. Details on the battery of biomarkers used in ALGNM

Organisms	Biomarkers
Fish	Metallothioneins level (E), Ethoxyresorufin O-deethylase (EROD) activity (E), Acetylcholinesterase activity (E), Vitellogenins level (E), Oxidative stress (S), DNA damage (G)
Bivalves	Metallothioneins level (E), Acetylcholinesterase activity (E), AcylcoA oxidase activity (E), Lysosome membrane stability (S), DNA damage (G), Oxidative stress (S), Lipid peroxidation products (S)

S: biomarkers of stress; G: biomarkers of genotoxicity; E: biomarkers of exposure.

Bioassays

The toxicity tests are used as a complement to chemical measures for quantifying the potential toxic effects, together with the bioavailability of chemical substances and the possible synergies [29 and therein]. Additionally, they also assess effects of those contaminants that have not been detected chemically [30]. A corrected toxicity assessment requires employment of species belonging to several taxa and main ecological or trophic positions [31 and therein].

ALNGM provides the use of a decomposer (*Vibrio fischeri*, Bacteria), a producer (*Dunaliella tetiolecta,* Algae) and two consumers (*Brachionus plicatilis*, Rotifera or *Tigriopus fulvus*, Crustacea and *Paracentrotus lividus*, Echinodermata). This bioassay battery allows observation of both sub-lethal (bacterial bioluminescence, algal growth, fertilization and development rates with sea urchins) and lethal (rotifer or crustacean mortality) endpoints. Toxicity tests are applied both on waters and sediments (marine and brackish samples). Water toxicity tests allow us to evaluate effects correlated to hydrophilic contaminants and particulate matter that are in the water column. Instead, sediment ecotoxicity allows assessment of toxic effects of absorbed contaminants that are bioavailable for aquatic organisms. All organisms are exposed directly to seawater samples; only *V. fischeri* is exposed to whole sediment while the other species are exposed to elutriate obtained by sediment [32].

Bioassays are carried out according to standardized protocols and data are elaborated by statistical programs, to analyze the relationship between dose and response; effect concentration or percentage is calculated and a toxicity judgment is given.

Ecological Analysis

Planktonic Communities

The planktonic community can provide no immediate answers but they can gradually, over time, provide information with regard to long-term consequences on the higher levels of the pelagic food web. Waters used in the re-gasification process are subjected to chlorination, undergoing changes in temperature and pressure, with the consequent possible death of the planktonic organisms inhabiting them. The subsequent release of water into the sea with chlorine content can also produce effects on planktonic communities living around the discharge area, in terms of diversity, functionality and reproduction, which must be evaluated both in the short and long term. The release of cooler water, in relation to the discharge depth and prevailing local circulation, can affect the conditions of mixing, with an exchange between water masses otherwise rich from a trophic point of view, which may affect the planktonic productivity in terms of biomass and biodiversity. The study of planktonic communities is conducted both qualitatively (taxa composition) and quantitatively (biomass) on phytoplankton, zooplankton and *ichthyoplankton*.

Fish Assemblages

Monitoring of fish assemblages is intended to highlight any changes in the structure of populations due to the construction and operation phase. These modifications may derive a) from physical-chemical disruption caused by the construction phase b) by the presence of Terminal GBS during the operational phase, and c) by the macrovacuolar barrier placed around the terminal as erosion protection and the artificial reef. The fish assemblage study was developed in the whole area around the terminal (macroscale study) and in the proximity of the structure (microscale study) in order to assess eventual changes of communities due to terminal presence and/or related activities. Qualitative investigations on the fish community living in the area are performed around the terminal macrovacuolar barrier, "*tegnùe*" and artificial reef by indirect methodologies (ROV), while quantitative investigations are performed by fishing with professional towed and fixed gears in the terminal area and artificial reef areas.

Soft-Bottom Zoobenthic Community

Benthic communities represent the biological memory of the ecosystem and thus provide information about previous perturbations. Analysis of benthic assemblage structure, both macro and meio fauna, is thus of paramount importance in each of the three phases identified in the ALNGM, in order to characterize environmental conditions of the study area and then to assess the potential environmental impacts deriving from the offshore terminal and its pipeline. The construction and production of the terminal can induce changes in the characteristics of sediment near the structures: the extent to which the biota is affected by these types of perturbations depends on complex interactions between environmental characteristics of the site and the characteristics of the installations. The terminal and pipeline construction might have some sort of impact on biological communities inhabiting the surrounding seabed [10]. For these reasons, qualitative and quantitative analyses are performed and ecological studies carried out. As the macrofauna is concerned, the systematic recognition of organisms is made at the species level at least for the most representative groups (e.g. Polychaeta, Mollusca, Crustacea, Echinodermata); for meiofauna at the order-phylum level (e.g. Nematoda, Copepoda, Polycheta, Mollusca, Gastrotricha). The number of individuals is counted for each taxa. The data thus obtained must be used for the calculation of the following biological indices:

- Abundance;
- Density;
- Total taxon richness;
- Indices of diversity, e.g. Shannon-Weaver H' [33].
- Indices of evenness, e.g. J' [34].

Multivariate analysis (e.g. Anova, Cluster Analysis, non-Multi Dimensional Scaling, etc.) and appropriate indices to assess the stress level of the communities (e.g. index M-AMBI – Multivariate AZTI Biotic Index) [35, 36] should be applied.

Hard-Bottom Zoobenthic Community

The terminal GBS, together with the macrovacuolar barrier and the artificial reef, on the soft-bottom seabed, can act as artificial hard substrates, where the establishment of benthic organisms is possible, exercising attraction to nektonic species. The study of macrobenthic assemblage and the formation and evolution of its settlement on a "virgin" substrate, then, is important to assess the real changes induced on the marine area by the existence of the structure. The study of hard-bottom zoobenthic community is conducted around the terminal (the macrovacuolar barrier) and the artificial reef, through indirect methodologies (photographic, ROV surveys), while close to "*tegnùe*" also through direct methodologies (suction sampling devices operated by divers).

Bioacoustics

Bioacoustics are planned in order to eventually observe noises due to the plant's construction and operation activities that could affect marine organisms living in the area; also the maritime traffic close to GBS was considered. In fact noises, related to anthropogenic activities in the water, could have a negative impact on several groups of animals; in particular marine turtles, mammals and fishes [37].

CONCLUSION

The LNG regasification Terminal located in the Northern Adriatic Sea is the first offshore LNG terminal in Italy and the first Gravity Based Structure in the world for unloading, storing and regasifying.

The ALNGM is the first Italian Monitoring Plan designed in order to assess potential environmental impacts linked to the typology of this project.

In the framework of an ecosystem approach, as recommended by the latest most innovative European legislations in the field of environmental protection, this Plan can be taken as a valuable reference for further studies and a useful tool for implementation of *ad hoc* monitoring plans.

The ALNGM is devised as a flexible tool allowing modifications when the activities are in progress, in relation with the acquired results.

Since it was born as an integrated plan, all outcomes obtained contribute to depict, at well as possible, the emerging environmental framework.

REFERENCES

[1] Edison (2011). How does the regasification process work? http://www.edison.it/en/media/press-releases/

[2] Edison (2009). First LNG cargo arrives at Adriatic LNG Terminal http://www.edison.it/en/media/press-releases/

[3] Ministry for the Environment and the Protection of Land and Sea, 1999, decree No. DEC/VIA/1999/4407.

[4] Ministry for the Environment and the Protection of Land and Sea, 2007, decree No. DEC/DSA/2007/975.

[5] Ministry for the Environment and the Protection of Land and Sea, 2004, decree No. DEC/DSA/2004/866.

[6] Kamp, R. (2005). An Overview of environmental and other impacts of the potential DKRW LNG plant/pipeline and Arizona Clean Fuels refinery in Tacna, Arizona with oil port and pipeline. Report for Proyecto Fronterizo de Educacion Ambiental, Playas de Tijuana, Baja California August 19, pp.1-26 www.etechinternational.org.

[7] Legislative Decree No.152 of 3 April 2006, GU. No. 88 of 14 April 2006.

[8] ICRAM (2005). Piano di Monitoraggio ambientale per il progetto di realizzazione ed esercizio del Terminale marino di rigassificazione LNG nel Nord Adriatico.

[9] Malacic, V.; Faganeli, J. and Malej, A. 2008. In Integration of Information for Environmental Security; Gonca Coskun, H. et al.; Eds.; Springer, 2008; pp 375-395.

[10] Chan, A.; Hartline, J.; Hueley, R.; Sruzziery, L. (2004). Evaluation of Liquefied Natural Gas receiving terminals for Southern California. MESM Group Project Report. University of California, Santa Barbara, pp. 1-138.

[11] European Parliament and European Council Directive 2000/60/EC of 23 October 2000.

[12] Chapman, P. M. *Sci. Total Environ.* 1990, vol 97, No 8, 815-825.

[13] Chapman, P. M.; Paine, M.D.; Arthur, A.D.; Taylor, L.A. *Mar. Pollut. Bull.* 1996, vol. 32, No 1, 47-64.
[14] Ramsar Convention (1971). UN Treaty Series No. 14583. Ramsar (Iran), 2 February 1971.
[15] European Council Directive 92/43/EEC of 21 May 1992.
[16] ASTM E1391-3:2008 Standard Guide for Collection, Storage, Characterization, and Manipulation of Sediments for Toxicological Testing and for Selection of Samplers Used to Collect Benthic Invertebrates.
[17] Ellis, J. I. and Schneider, D. C. *Environ. Monit. and Assess.* 1997, vol 48, 157-172.
[18] Cicero, A. M.; Gabellini, M. ; Maggi, C. ; Nonnis, O.; Trabucco, B. and Virno Lamberti, C. Rapp. *Comm. Intern. Mer Medit.* 2004, vol 37, 507.
[19] Sea Bird Electronics (2007). Sbe 19 plus User's manual. Sea Bird Electronics Inc. Washington: Bellevue.
[20] Baumgartner, D. J.; Frick W. E.; Roberts P. J. W. 1994. In Dilution models for effluent discharges. Washington DC: Environmental Protection Agency Report No. EPA/600/R-94/086.
[21] Ministry for the Environment and the Protection of Land and Sea, decree of 24 January 1996, GU. No. 31 of 7 February 1996
[22] Ministry for the Environment and the Protection of Land and Sea, decree No. 56 of 14 April 2009.
[23] European Parliament and European Council Decision No. 2455/2001/EC of 20 November 2001, GU. No. L 331 of 15 December 2001.
[24] Kenny, A. J.; Cato, I.; Desprez, M.; Fader, G., Schuttenhelm, R. T. E. and Side, J. ICES *J. of Mar. Sci.,* 2003, vol 60, No 2, 411- 418.
[25] Collier, J. S. and Brown, C. *J. Mar. Geol.*, 2005, vol 214, 431- 449.
[26] Depledge, M. H. In Nondestructive Biomarkers in Vertebrates; Fossi, M. C., Leonzio, C.; Eds.; Lewis: Boca Raton, FL, 1993, 261-285.
[27] Viarengo, A.; Lowe, D.; Bolognesi, C.; Fabbri, E.; Kohler, A. *Comparative Biochemistry and Physiology*, 2007, 146 part C, 281-300.
[28] Bocchetti, R.; Virno Lamberti, C.; Pisanelli, B.; Razzetti, E. M.; Maggi, C.; Catalano, B.; Sesta, G.; Martuccio, G.; Gabellini, M.; Regoli, F. *Mar. Environ. Res.* 2008, 66, 24-26.
[29] Manfra, L. and Maggi, C. In *Advances in Natural Gas Technology*; ISBN 979-953-307-567-8 InTech 2012, 271-284.
[30] Ahlf, V. and Heise, S. *J. Soils and Sediment.*, 2005, vol 5, No 1, 16-20.
[31] Piva, F.; Ciaprini, F.; Onorati, F.; Benedetti, M.; Fattorini, D.; Ausili, A. and Regoli, F. *Chemosphere,* 2009, vol 83, 475-485.

[32] USEPA (1991). Technical support document for water quality-based toxic control. 505/2-90-001, US Environmental Protection Agency.

[33] Shannon, C.E. and Weaver, W. 1949. The mathematical theory of communication. Urbana IL: University of Illinois Press. 117 pp.

[34] Pielou, E.C. Population and community ecology, principles and methods. *Gordon and Breach Sci. Pubbl.*: New York, 1974; 424 pp.

[35] Muxika, I.; Borja, A. and Bald, *J. Mar. Pollut. Bull.* 2007, vol 55, 16-29.

[36] Clarke, K. R.; Gorley, R. N. (2006). PRIMER v6: User Manual/Tutorial. PRIMER-E: Plymouth.

[37] André, M.; Solé, M.; Lenoir, M.; Durfort, M.; Quero, C.; Mas, A.; Lombarte, A.; van der Schaar, M.; López-Bejar, M.; Morell, M.; Zaugg S. and Houégnigan L. *Front. Ecol. Environ.* 2011, vol 9, No 9, doi:10.1890/100124.

In: Mediterranean Sea
Editor: Terrence B. Hughes
ISBN: 978-1-62618-238-7

Chapter 6

BIOACTIVE PEPTIDES FROM THE VENOM OF THE MEDITERRANEAN CONE SNAIL *CONUS VENTRICOSUS*

Maria Carolina Spiezia[1], Cristiano Chiarabelli[1], Maria Eugenia Schininà[2] and Fabio Polticelli[1,3,*]

[1]Department of Biology, University of Roma Tre, Viale Guglielmo Marconi, Rome, Italy

[2]Department of Biochemical Sciences, Sapienza University of Roma, Piazzale Aldo Moro,Rome, Italy

[3]National Institute of Nuclear Physics, Roma Tre Section, Via della Vasca Navale, Rome, Italy

ABSTRACT

Cone snails are marine predator gastropods that represent an incredible source of bioactive peptides also known as conopeptides. Their venom is a complex mixture of different types of (macro)molecules but in all the species the most abundant component is represented by conotoxins, disulphide-rich peptides that display the cysteine knot structural motif. Because of their high specificity and potency to bind and inhibit ion channels and receptors, conotoxins represent powerful tools

* Corresponding author: Fabio Polticelli, Department of Biology, University of Roma Tre, Viale Guglielmo Marconi 446, I-00146 Rome, Italy. Phone: +39-06-57336362; Fax: +39-06-57336321; E-mail: polticel@uniroma3.it.

for neuroscience as well as lead compounds for drug discovery. *Conus ventricosus*, a worm-hunting cone snail, is the single cone snail species living in the Mediterranean Sea and the characterization of its venom started about ten years ago. The first *C. ventricosus* peptide isolated and characterized was Contryphan-Vn, a 9 amino acids peptide which displays activity on both invertebrate and vertebrate Ca^{2+} dependent K^{+} channels. Later, *C. ventricosus* venom peptides profiling led to the discovery of an additional conopeptide of the Leu-Contryphan family, whose molecular target is still unknown, and two novel conotoxins, named α-Conotoxin-Vn and Conotoxin-Vn2. The former one targets nicotinic acetylcholine receptors and the latter most probably voltage-gated Ca^{2+} channels. Recently a recombinant expression system has been developed to increase the yield in soluble Conotoxin-Vn2 and perform structural and functional analysis. This analysis revealed that conotoxin-Vn2 displays a high toxic activity towards insects larvae, providing new means for the development of a novel generation of environmental friendly pesticides.

Keywords: Cone snails, Conopeptides, Conotoxins, Ion Channels

INTRODUCTION

Conopeptides are bioactive peptides from the venom of mollusks of the genus *Conus*, widespread in tropical and subtropical marine environment. Even though cone snails venom is a complex mixture of different types of (macro)molecules (Craig et al., 1999) which act in synergy to immobilize prey, conopeptides represent the most abundant component (Olivera et al., 1991). All conopeptides are initially expressed as large precursor polypeptides that undergo post-translational modifications to yield the mature peptide, generally 10 to 40 amino acids long (Buczek et al., 2005). While the pre- and pro-region of precursor peptides are well conserved, the mature peptides are characterized by a high sequence variability and their structure is stabilized by an unusual number of conserved disulphide bonds. This is especially true for conotoxins, conopeptides that display the presence of the cysteine knot structural motif in which four cysteine residues form two knotted disulphide bonds (Craik et al., 2001). Conticello and coworkers (2001) studying conopeptide precursor alignments observed an accelerated rate of nucleotide substitution in the mature peptide region with respect to the pre- and pro-regions and a position-specific conservation of cysteine codons within the hypervariable region. To explain the evolutionary origin of conopeptide sequences hypervariability, they thus proposed a mutator mechanism targeted to mature domains in conopeptide genes, combining a protective activity specific for cysteine codons and a mutagenic polymerase that exhibits transversion bias (Conticello et al., 2001). More recently it has been proposed that several evolutionary mechanisms might be acting concomitantly such as genes duplication and a high rate of non-synonymous substitutions (Duda and Remigio, 2008).

About 500 cone snail species are known, and it has been estimated that there are about 100,000 peptides, each one with peculiar pharmacological properties (Terlau and Olivera, 2004). Despite their large number, all conopeptides are produced by translating mRNA from genes belonging to only sixteen superfamilies divided according to the relative conservation of the signal peptide (Kaas et al., 2010). They are also divided in groups based on the cysteine framework, and in pharmacological families based on their molecular target (Kaas et al., 2010). Among conopeptides, the most abundant components of cone snails venom is represented by conotoxins, characterized by at least two knotted disulphide bonds (see Kaas et al., 2010 and references therein). Because of their high potency and target specificity, conotoxins have attracted much interest as tools for investigating ion channel function, as neuropharmacological tools, as potential leads in drugs development, or indeed as drugs themselves (Adams et al., 1998). Currently the synthetic analogue of ω-conotoxin MVIIA from *Conus magus*, named Ziconotide, is used as an analgesic drug under the commercial name of Prialt (Wermeling, 2005).

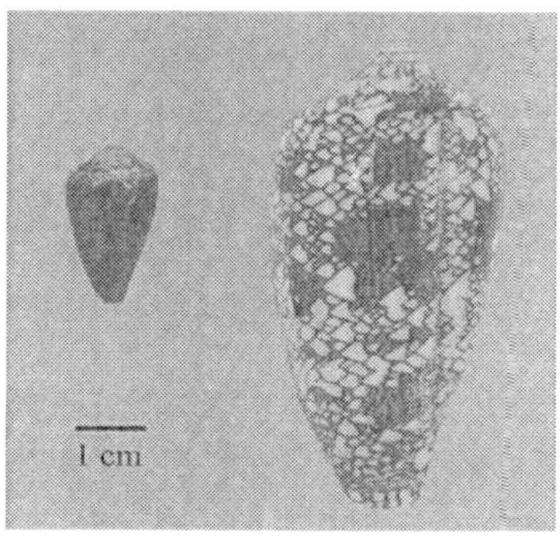

Figure 1. Specimens of the worm-hunting *Conus ventricosus* (left) and of the mollusk-hunting *Conus textile* (right).

Conus ventricosus, a worm-hunting cone snail, is the single cone snail species living in the Mediterranean Sea. Due to their small size, *Conus ventricosus* specimens are generally 2-3 cm in length (Figure 1), and the small amount of venom that can be extracted from their venom duct, the characterization of *C. ventricosus* venom components has been difficult. However, in the last few years several *C. ventricosus* conopeptides have been isolated and characterized revealing interesting structural and functional properties (Raybaudi Massilia et al., 2001; 2003; Eliseo et al., 2004; Romeo et al., 2008, Spiezia et al., 2012). Here the current knowledge on *C. ventricosus* venom components is reviewed.

Intraspecific Venom Variability

The Mediterranean Sea provides different trophic habitats for the *C. ventricosus* cone snails, thus each *C. ventricosus* population develops a unique conopeptides repertoire, depending on its natural habitat. With the aim to compare the venom characteristics of individual populations within *C. ventricosus* species, the venom composition of several specimens collected in three different Mediterranean Sea regions was analyzed: Tyrrhenian Sea (Civitavecchia, Rome), Ionian Sea (Porto Selvaggio, Lecce) and southern Mediterranean (Lampedusa, Pelagie islands). Extraction of the venom from venom ducts was followed by MALDI-TOF mass spectrometry analysis on peptide mixtures revealing significant differences in the venom composition (Romeo et al., 2008). The specimens collected in Civitavecchia showed the most heterogeneous peptide profile comprising peptides with a broad range of molecular weights, while the Porto Selvaggio and Lampedusa ones revealed a common and peculiar composition, characterized by a more limited number of peptides (Romeo et al., 2008). Such a peptides diversification might be due to the different selective pressure of the three groups of specimens to interact with prey and predators in their natural habitats. Combining HPLC and MALDI-TOF analysis it was possible to determine mass matches to specific conopeptides classes for a number of *C. ventricosus* peptides (Table 1). MALDI-TOF analysis and cysteinil derivatization were used to obtain structural information on the most abundant conopeptides identified, by determining the number of disulphide bridges. These results, together with direct Edman degradation on selected HPLC purified peptides, allowed the determination of the amino acid sequences of these novel conopeptides, and hence their cysteine framework (Table 2), an additional characteristics which allows classification of conopeptides (Kaas et al., 2010).

Table 1. Summary of the most abundant *C. ventricosus* conopeptides

Monoisotopic mass (Da)	Site of collection	Proposed conopeptide family	Complete amino acid sequence
1030.4	Civitavecchia	Contryphan	$DCPWKPWC_{NH2}$
1073.4	Civitavecchia	Contryphan	$GDCPWKPWC_{NH2}$
1087.5	Civitavecchia	Contryphan	
1179.3	Lampedusa, Porto Selvaggio	Contryphan	
1716.7	Lampedusa, Porto Selvaggio	Conotoxin	$GGCCSYPPCAVSNPQHC_{NH2}$
1740.9	Lampedusa, Porto Selvaggio	Conotoxin	
1742.9	Lampedusa, Porto Selvaggio	Conotoxin	
1991.5	Civitavecchia	Conotoxin	
2927.8	Civitavecchia	Conotoxin	
3144.1	Lampedusa, Porto Selvaggio	Conotoxin	
3432.5	Lampedusa, Porto Selvaggio	Conotoxin	
3536.7	Civitavecchia	Conotoxin	$EDCIAVGQLCVFWNIGRPCCSGLVFAC_{NH2}$

Adapted from Romeo et al., 2008.

Table 2. ***C. ventricosus*** **conopeptides so far characterized. The cysteine framework and the putative physiological target are indicated**

Name	Proposed superfamily/family	Disulphide framework	Pharmacological target	Ref.
Leu-Contriphan	Contryphan	C-C	Unknown	This work
Contryphan-Vn	Contryphan	C-C	Ca^{2+}-dependent K^{+} channels	Raybaudi Massilia et al., 2003
Conotoxin-Vn2	O/ω Conotoxin	C-C-CC-C-C	Ca^{2+} channels	Romeo et al., 2008
α-Conotoxin-Vn	A/α Conotoxin	CC-C-C	AChR receptor	Romeo et al., 2008

Adapted from Romeo et al., 2008.

Table 3. Amino acid sequence alignment of selected Contryphansa

Peptide	Sequence	Species	Prey	Ref.
Contryphan-R	G—-CO-W̲EPWC*	*C. radiatus*	fish	(Jacobsen et al., 1998)
Contryphan-R/Tx	G—-CO-W̲EPWC*	*C. textile*	mollusk	(Jimenez et al., 2001)
Contryphan-P	G—-CO-W̲DPWC*	*C. purpurascens*	fish	(Jimenez et al., 1996)
Bromocontryphan	G—-CO-W̲EP~~W~~C*	*C. radiatus*	fish	(Jimenez et al., 1997)
Contryphan-Sm	G—-CO-W̲QPWC*	*C. stercusmuscarum*	fish	(Jimenez et al., 1996)
Contryphan-Tx	G—-CO-W̲QPYC*	*C. textile*	mollusk	(Jimenez et al., 2001)
Contryphan-Vn	GD-CP-W̲KPWC*	*C. ventricosus*	worm	(Raybaudi Massilia et al., 2001)
Leu-contryphan-Tx	---CV-L̲YPWC	*C. textile*	mollusk	(Jimenez et al., 2001)
Leu-contryphan-P	G—-CV-L̲LPWC	*C. purpurascens*	fish	(Jacobsen et al., 1999)
Leu-contryphan-Vn	GDFCVSL̲-PWC*	*C. ventricosus*	worm	(Unpublished data)

[a] O indicates hydroxyproline, L̲ D-leucine, W̲ D -tryptophan, ~~W~~ brominated tryptophan and * *C-terminus* amidation. Amino acid substitutions with respect to Contryphan R are in bold-type.

C. Ventricosus Conopeptides of the Contryphan Family

Contryphans are small cyclic bioactive peptides, first isolated from the venom of fish-hunting and mollusk-hunting cone snails, characterized by a huge number of unusual post-translational modifications (see Raybaudi Massilia et al., 2001 and references therein). Contryphan-Vn (Swiss Prot accession no. P83047.1), a 9 amino acids peptide, was the first bioactive peptide purified and characterized from the *C. ventricosus* venom (Raybaudi Massilia et al., 2001). Mass spectrometry analysis on Contryphan-Vn revealed the N-terminal sequence common to the Contryphan family, and confirmed the C-terminal amidation and the presence of a single disulphide bridge in the native peptide, as observed in all the contryphans previously characterized (Jimenez et al., 1996; Jimenez et al., 1997; Jacobsen et al., 1998; Jimenez et al., 2001; Raybaudi Massilia et al., 2001). Direct Edman degradation coupled with MS/MS analysis showed that Contryphan-Vn displays a high sequence similarity with eight Contryphan peptides from Indo-Pacific fish-hunting and mollusk-hunting species as well as some peculiar features (Figure 2). In particular, HPLC coelution experiments of the natural peptide with L- and D-Trp8 synthetic variants demonstrated the presence of a D-Trp residue in the natural peptide, characteristic conserved in all Contryphans (Raybaudi Massilia et al., 2001). On the other hand, differently from all the other contryphans, Contryphan-Vn displays the unusual presence of the Lys6 residue in the intercysteine loop that conferred a positively charged surface to the peptide. The solution structure of Contryphan-Vn (Protein Data Bank code 1NXN; Eliseo et al., 2004) obtained by NMR-spectroscopy revealed the presence of two β-turn elements, and the involvement of the Lys6 is a salt bridge stabilizing the *cis* peptide isomer rather than the *trans* one (Figure 2). On the basis of homology modeling studies (Raybaudi Massilia et al., 2001), it had been previously proposed that in Contryphan-Vn Lys6 and Trp8 could form a dyad reminiscent of the Lys-aromatic dyad characteristic of several K^+ channel blocker toxins (Dauplais et al., 1997), suggesting K^+ channels as potential targets for Contryphan-Vn.

In order to verify this hypothesis, electrophysiological experiments on dorsal unpaired median (DUM) neurons from cockroach nerve cord and on rat fetal chromaffin cells, were carried out (Raybaudi Massilia et al., 2003). In the invertebrate system (DUM neurons), Contryphan-Vn exerted a complex action on K^+ channels dependent on Ca^{2+} influx through voltage-activated Ca^{2+}

channels. In fact the K^+ global current was inhibited in a manner similar to some scorpion toxins, but Contryphan-Vn caused an increase of the K^+ current amplitude when the voltage-activated Ca^{2+} channels were blocked (Raybaudi Massilia et al., 2003). Differently, in rat fetal chromaffin cells Contryphan-Vn elicited an effect on the K^+ outward current dependent on the release of Ca^{2+} from internal storage, acting on Ca^{2+}-dependent K^+ channels (Raybaudi Massilia et al., 2003). The voltage-gated and Ca^{2+}-dependent K^+ channels have represented the first functional target identified for a member of the Contryphan family, and Contryphan-Vn has been the first conopeptide known to modulate these channels types.

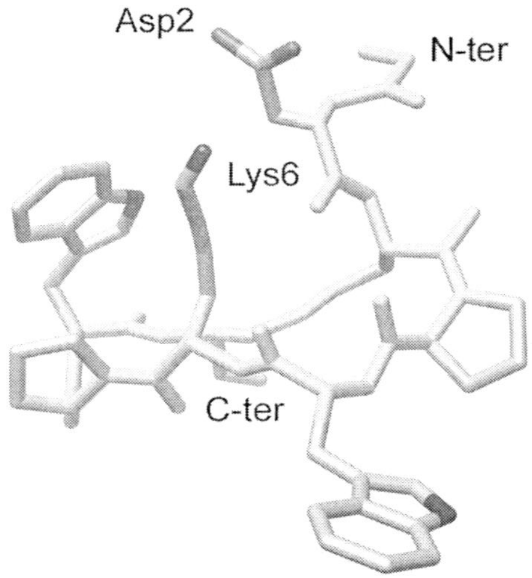

Figure 2. Solution structure of Contryphan-Vn obtained by NMR spectroscopy (Eliseo et al., 2004). Backbone atoms and cysteine residues are coloured in yellow.

Another variant of the Contryphan family of peptides is represented by Leu-Contryphans which display the presence of a D-Leu residue in place of D-Trp (Jimenez et al., 2001; Jacobsen et al., 1999). A peptide belonging to this family was also isolated from *C. ventricosus* venom. Edman degradation of the isolated peptide allowed the determination of its amino acid sequence (Table 3; unpublished data), showing interesting differences with respect to the other known peptides of this family, Leu-Contryphan-Tx and Leu-Contryphan-P (Jimenez et al., 2001; Jacobsen et al., 1999) However, as for other Leu-Contryphans, the molecular target of Leu-Contryphan-Vn is still unknown.

C. Ventricosus Conotoxins

The mass spectrometric analysis of conopeptides profiles on specimens from different Mediterranean Sea sites coupled to cysteine reduction and alkylation (see above), revealed the presence of several novel conotoxins. Direct Edman degradation on the most abundant HPLC purified peptides led to the identification of two novel conotoxins, named α-Conotoxin-Vn (SwissProt accession no. P0C8V4.1) and Conotoxin-Vn2 (SwissProt accession no. P83301.1) (Table 2; Romeo et al., 2008). On the basis of the common disulphide pattern (-CC-C-C-) and the high degree of homology with the primary structures of known α-conotoxins, the α-conotoxin-Vn was indeed assigned to the α/A family, which comprises antagonists of neuronal and muscular nicotinic acetylcholine receptors (nAChRs). Analysis of α-conotoxin-Vn molecular model, built using as a template the homologous A10L-D14K variant of α-Conotoxin PnIA (PDB code: 2BR8; Celie et al., 2005), supported this hypothesis. The latter conotoxin variant had been crystallized in complex with the acetylcholine-binding protein (AChBP) structure, the prototype of the acetylcholine-binding domain of the nAChRs (PDB code: 2BR8; Celie et al., 2005). Thus, to test the hypothesis that α-Conotoxin-Vn binds to the same molecular target of α-Conotoxin PnIA (Hu et al., 1996; Celie et al., 2005), the putative complex formed by α-Conotoxin-Vn and AchBP was also modeled (Romeo et al., 2008). Analysis of the modeled complex evidenced the conservation of the main molecular interactions determining the binding of α-Conotoxins to the nicotinic acetylcholine receptors, establishing that residues Pro8, Pro11, Ser12 and Gln15 of α-Conotoxin-Vn (Figure 3) play the same structural role of residues Pro7, Leu10, Asn11, and Lys14 of the A10L-D14K variant of α-Conotoxin PnIA (Romeo et al., 2008).

Analysis of Conotoxin-Vn2 amino acid sequence revealed a cysteine framework common to I1, I3, M, O1, O2 and O3 conotoxin gene superfamilies (Kaas et al., 2010), characterized by the common cysteine residues pattern -C-C-CC-C-C (Table 2). Amino acid sequence alignment between all disulphide rich bioactive peptides with known three-dimensional structure sharing this cysteine framework revealed the highest degree of similarity of Conotoxin-Vn2 with δ-conotoxin EVIA, a peptide extracted from the venom of *Conus ermineus* that selectively acts on vertebrate neuronal Na^+ channels (Volpon et al., 2004), and the Ptu1 toxin, a toxin from the assassin bug *Peirates turpis* that blocks the N-type voltage-sensitive Ca^{2+} channels (Bernard et al., 2001). Despite several attempts, the extreme hydrophobicity of Conotoxin-Vn2

(Table 1) did not allow to obtain by peptide synthesis a sufficient amount of the peptide to carry out bioactivity assays to identify the molecular target of this toxin. A structural model of Conotoxin-Vn2 was thus built by homology modeling, using as templates the available structures of δ-conotoxin EVIA and Ptu1 toxin (PDB codes 1G1Z and 1I26, respectively), in the attempt to identify a known pharmacophore.

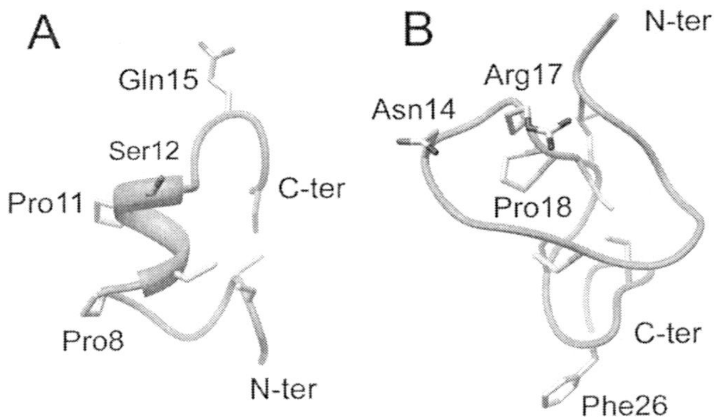

Figure 3. Schematic representation of the molecular models of *Conus ventricosus* α-Conotoxin-Vn (A) and Conotoxin-Vn2 (B) (Romeo et al., 2008). Residues involved in target recognition are shown in stick representation, coloured by atom type and labeled (see text). Cysteine residues are also shown and coloured in yellow.

Analysis of the molecular model of Conotoxin-Vn2 revealed an amphipatic molecular surface, with hydrophobic residues (Leu^9, Val^{11}, Phe^{12}, Trp^{13}, Leu^{23}, Val^{25}, Phe^{26}) on one side and acidic (Glu^1 and Asp^2) and basic (Arg^{17} e Lys^{33}) residues on the opposite side (Romeo et al., 2008). Interestingly, comparison of the Conotoxin-Vn2 structural model with the available structures of voltage-dependent calcium channels toxins, revealed the presence of an array of residues similar to the pharmacophore present in ω-atracotoxin Hvla, a spider neurotoxin acting on insects high voltage activated Ca^{2+} channels (Tedford et al., 2004). In particular, residues Pro10, Tyr13, Asn27 and Arg35 of spider ω-atracotoxin Hvla, known to be responsible for binding of this toxin to its molecular target, display a spatial arrangement similar to Conotoxin-Vn residues Pro18, Phe26, Asn14 and Arg17 (Table 1 and Figure 3).

Given the above mentioned difficulties in chemical synthesis of Conotoxin-Vn2, to get a deeper insight on the bioactivity of this peptide, a system has been recently developed to express it in recombinant form in *E. coli* cells as a GST-fusion product (Spiezia et al., 2012). Since the natural prey of *C. ventricosus* is represented by invertebrate polychaete worms, the GST-Conotoxin-Vn2 was assayed on a well known invertebrate model system, the larvae of the moth *Galleria mellonella*. In this system Conotoxin-Vn2 displayed strong insecticidal properties at a dose of only 100 pmol/g of body weight (Spiezia et al., 2012). Moreover, because a negatively charged residue immediately before the first cysteine residue is conserved in all the available amino acid sequences of *C. ventricosus* ω-like conotoxins, the Asp2His mutant was also produced in order to analyze the role of this Asp residue in the toxin bioactivity. Contrary to expectations, the Asp2His mutant displayed enhanced toxicity in the larvae model system, ruling out the possibility that the conserved acidic residue is involved in the recognition of Conotoxin-Vn2 biological target, at least in the insect model system (Spiezia et al., 2012).

Interestingly, Conotoxin-Vn2 proved to be highly toxic also in a vertebrate model system, the fish *Gobius niger*, a fish of the Gobidae family which shares the same coastal habitat with *C. ventricosus*. In fact, intraperitoneal injection of the Conotoxin-Vn2 at a dose of only 100 pmol/g of body weight caused frenzy swimming behaviour of the fish, followed by tetanic paralysis and death, while lower doses caused similar initial symptoms followed by recovery of the fish (Spiezia et al., 2012). This result demonstrates that *C. ventricosus* venom can potentially be used by this snail not only to immobilize preys but also as a defense against predators.

CONCLUSION

Conus ventricosus is the single cone snail species living in the Mediterranean Sea. Given its small dimensions and worm-hunting habit, it has been neglected for a long time by scientist working in the toxinology field with respect to the more studied mollusk- and fish-hunting tropical species. However, the characterization of the *C. ventricosus* venom components led to the discovery of several novel conopeptides with interesting pharmacological properties. In particular, Contryphan-Vn was the first peptide of this family for which a functional target was identified (Raybaudi Massilia et al., 2003). Further, the characterization of both Contryphan-Vn and Conotoxin-Vn2 GST fusion protein demonstrated that *C. ventricosus* conopeptides are active not only on invertebrate but also in vertebrates model systems (Raybaudi Massilia et al., 2003; Spiezia et al., 2012). This finding confirms that even vermivorous cone snails are probably capable of defending themselves from predators using their venom.

Finally, the property of *C. ventricosus* conopeptides to target insects ion channels and receptors, as demonstrated both for Contryphan-Vn and Conotoxin-Vn2, makes them attracting tools for the development of a novel generation of environmental friendly pesticides. Indeed, expression in tobacco plants of the wild type and Asp2His variants of Conotoxin-Vn2 is in progress in order to evaluate the possibility to develop conotoxins-based insect-resistant transgenic crops by challenging the transgenic plants with phytophagous *Manduca sexta* (Tobacco Hornworm) larvae.

REFERENCES

Adams, D.J., Alewood, P.F., Craik, D.J., Drinkwater, R.D., Lewis, R.J., (1999). Conotoxins and their potential pharmaceutical applications. *Drug Develop Res*, 46:219-234.

Bernard, C., Corzo, G., Mosbah, A., Nakajima, T., Darbon, H., 2001. Solution structure of Ptu1, a toxin from the assassin bug *Peirates turpis* that blocks the voltage-sensitive calcium channel N-type. *Biochemistry*, 40:12795-12800.

Buczek, O., Bulaj, G., Olivera, B.M., 2005. Conotoxins and the posttranslational modification of secreted gene products. *Cell. Mol. Life Sci.*, 62:3067-3079.

Celie, P.H., Kasheverov, I.E., Mordvintsev, D.Y., Hogg, R.C., van Nierop, P., van Elk, R., van Rossum-Fikkert, S.E., Zhmak, M.N., Bertrand, D., Tsetlin, V., Sixma, T.K., Smit, A.B. 2005. Crystal structure of nicotinic acetylcholine receptor homolog AChBP in complex with an alpha-conotoxin PnIA variant. *Nat. Struct. Mol. Biol.*, 12:582-588.

Conticello, S.G., Gilad, Y., Avidan, N., Ben-Asher E., Levi, Z., Fainzilber, M., 2001.. Mechanisms for evolving hypervariability: the case of conopeptides. *Mol. Biol. Evol.*, 18:120-131.

Craig, A.G., Norberg, T., Griffin, D., Hoeger, C., Akhtar, M., Schmidt, K., Low, W., Dykert, J., Richelson, E., Navarro, V., Mazella, J., Watkins, M., Hillyard, D., Imperial, J., Cruz, L.J., Olivera, B.M., 1999. Contulakin-G, an O-glycosilated invertebrated neurotensin. *J. Biol. Chem.*, 274:13752-13759.

Craik, D.J., Daly, N.L., Waine, C., 2001. The cystine knot motif in toxins and implications for drug design. *Toxicon*, 39:43-60.

Dauplais, M., Lecoq, A., Song, J., Cotton, J., Jamin, N., Gilquin, B., Roumestand, C., Vita, C., de Medeiros, C.L., Rowan, E.G., Harvey, A.L., Mènez. A., 1997. On the convergent evolution of animal toxins. Conservation of a dyad of functional residues in potassium channel-blocking toxins with unrelated structures. *J. Biol. Chem.*, 272:4302-4309.

Duda, T.F. Jr, Remigio, E.A., 2008. Variation and evolution of toxin gene expression patterns of six closely related venomous marine snails. *Mol. Ecol.*, 17:3018-3032.

Eliseo, T., Cicero, D.O., Romeo, C., Schininà, M.E., Raybaudi Massilia, G., Polticelli, F., Ascenzi, P., Paci, M., 2004. Solution structure of the cyclic peptide Contryphan-Vn, a Ca^{2+} dependent K^+ channel modulator. *Biopolymers*, 74:189-198.

Hu, S.H., Gehrmann, J., Guddat, L.W., Alewood, P.F., Craik, D.J., Martin, J.L., 1996. The 1.1 Å crystal structure of the neuronal acetylcholine receptor antagonist, alpha-conotoxin PnIA from *Conus pennaceus. Structure*, 4:417-23.

Jacobsen, R., Jimenez, E.C., Grilley, M., Watkins, M., Hillyard, D., Cruz, L.J., Olivera, B.M., 1998. The contryphans, a D-tryptophan-containing family of *Conus* peptides: interconversion between conformers. *J. Pept. Res.*, 51:173-179.

Jacobsen, R.B., Jimenez, E.C., De la Cruz, R.G., Gray, W.R., Cruz, L.J., Olivera, B.M., 1999. A novel D-leucine-containing *Conus* peptide: diverse conformational dynamics in the contryphan family. *J. Pept. Res.* 54:93-99.

Jimenez, E.C., Craig, A.G., Watkins, M., Hillyard, D.R., Gray, W.R., Gulyas, J., Rivier, J.E., Cruz, L.J., Olivera, B.M., 1997. Bromocontryphan: post-translational bromination of tryptophan. *Biochemistry* 36:989-994.

Jimenez, E.C., Olivera, B.M., Gray, W.R., Cruz, L.J., 1996. Contryphan is a D-tryptophan-containing Conus peptide. *J. Biol. Chem.*, 271:28002-28005.

Jimenez, E.C., Watkins, M., Juszczak, L.J., Cruz, L.J., Olivera, B.M., 2001. Contryphans from *Conus textile* venom ducts. *Toxicon*, 39:803-808.

Kaas, Q., Westermann, J.C., Craik, D.J., 2010. Conopeptide characterization and classifications: an analysis using ConoServer. *Toxicon*, 55:1491-1509.

Olivera, B.M., Rivier, J., Scott, J.K., Hillyard, D.R., Cruz, L.J., 1991. Conotoxins. *J.Biol Chem*, 266:22067-22070.

Raybaudi Massilia, G., Eliseo, T., Grolleau, F., Lapied, B, Barbier, J., Bournaud, R., Molgó, J., Cicero, D.O., Paci, M., Schininà, M.E., Ascenzi, P., Polticelli, F., 2003. Contryphan-Vn: a modulator of Ca^{2+}-dependent K^+ channels. *Biochem. Biophys. Res. Commun.*, 303:238-246.

Raybaudi Massilia, G., Schininà, M.E., Ascenzi, P., Polticelli, F., 2001. Contryphan-Vn: a novel bioactive peptide from the venom of the Mediterranean worm-hunting marine snail *Conus ventricosus*. *Biochem. Biophys. Res. Commun.*, 288:908-913.

Romeo, C., Di Francesco, L., Oliverio, M., Palazzo, P., Raybaudi Massilia, G., Ascenzi, P., Polticelli, F., Schininà, M.E., 2008. *Conus ventricosus* venom peptides profilig by HPLC-MS: a new insight in the intraspecific variation. *J. Sep. Sci.*, 31:488-498.

Spiezia, M.C., Chiarabelli, C., Polticelli, F., 2012. Recombinant expression and insecticidal properties of a *Conus ventricosus* conotoxin-GST fusion protein. *Toxicon*, 60:744-751.

Tedford, H.W., Gilles, N., Ménez, A., Doering, C.J., Zamponi, G.W., King, G.F., 2004. Scanning mutagenesis of omega-atracotoxin-Hv1a reveals a spatially restricted epitope that confers selective activity against insect calcium channels. *J. Biol. Chem.*, 279:44133-44140.

Terlau, H., Olivera, B.M., 2004. Conus venom: a rich source of novel ion channel-targeted peptides. *Physiol Rev*, 84:41-68.

Volpon, L., Lamthanh, H., Barbier, J., Gilles, N., Molgó, J., Ménez, A., Lancelin, J.M., 2004. NMR solution structures of delta-conotoxin EVIA from *Conus ermineus* that selectively acts on vertebrate neuronal Na^+ channels. *J. Biol. Chem.*, 279:21356-21366.

Wermeling, D.P., 2005. Ziconotide, an intrathecally administered N-type calcium channel antagonist for the treatment of chronic pain. *Pharmacotheraphy*, 25:1084-1094.

Index

A

access, 18, 30, 37
acetone, 44, 45, 46
acetylcholine, xi, 136, 143, 146, 147
acid, 44, 140, 143
acidic, 144, 145
adjustment, 105, 109, 112
Adriatic LNG Terminal, x, 115, 116, 117, 119, 131
adverse effects, 19
aerosols, 62
aging population, 2
agriculture, 14, 59
algae, 55, 95
Algeria, 116
alkylation, 143
amino acid(s), xi, 95, 136, 138, 139, 141, 142, 143, 145
ammonia, 67
amplitude, 142
analgesic, 137
ANOVA, 127
anoxia, 102
appetite, 24
aquaculture, 14, 113
aromatic hydrocarbons, 61, 62
Artificial Neural Networks, 110, 111
Asia, 2
aspiration, 71
assessment, 16, 27, 38, 68, 120, 127, 128
assets, 17, 19
atmosphere, 62, 67
atoms, 95, 142
authorities, 20, 103
autonomy, 100

B

background information, 126
bacteria, 95, 97, 121, 125
banks, 101
base, 133
behavioral change, 126
bending, 50
benefits, vii, 1, 4, 5, 6, 7, 8, 9, 10, 17, 20, 22, 24, 30
bias, 137
bioaccumulation, 59, 70, 119, 120, 122
bioassay, 127
bioavailability, 61, 125, 127
biodegradation, 67
biodiversity, viii, 3, 4, 20, 23, 26, 28, 32, 36, 103, 120, 128
bioindicators, 122
biological responses, 127
bioluminescence, 127
biomarkers, 120, 122, 127
biomass, 57, 119, 128
biosphere, 9

biotic, 4, 59, 119
body weight, 145
bonds, 96, 136, 137
breakdown, 95
breeding, 119
Broadband, 74
bromination, 147

C

Ca^{2+}, xi, 136, 140, 141, 143, 144, 147
cables, 125
calcium, 144, 146, 148
capillary, 45, 46
capsule, 44
carbohydrates, 95
carbon, 3, 19, 21, 23, 30, 32, 43, 55, 62, 63, 95, 125
carboxylic acid(s), 95
case studies, 14, 62
C-C, 140, 143
cell size, 49, 50
challenges, 32
channel blocker, 141
chemical(s), 59, 60, 62, 67, 68, 70, 74, 94, 95, 103, 118, 119, 121, 124, 125, 126, 127, 128, 144
chemical characteristics, 118, 124
Chile, 30
chlorination, 128
chlorine, 59, 117, 118, 128
chlorophyll, 93, 95, 97, 113, 122
chromatography, 45, 46
CIA, 2
circulation, 69, 128
civilization, 2
clarity, 111
classes, 53, 138
classification, 14, 29, 138
clay minerals, 52
cleaning, 42
climate, 21
climate change, 21
coal, 57
coastal ecosystems, 8, 14, 17
coastal management, 14
codes, 37
codes of conduct, 37
collaboration, 4, 19, 25
collisions, 38
colonization, 24
color, 94
combustion, 57
combustion processes, 57
commercial, viii, 2, 8, 14, 36, 37, 38, 100, 117, 137
communication, 100, 133
community(s), 2, 4, 23, 61, 92, 118, 120, 122, 123, 128, 129, 130, 133
compatibility, x, 116
compensation, 17
competitiveness, 23
compilation, 11
complement, 127
complex interactions, 129
complexity, 5, 24, 93
compliance, 125
composition, 44, 52, 62, 63, 95, 96, 97, 112, 118, 128, 138
compounds, x, 59, 60, 67, 94, 95, 96, 125, 136
computation, 40, 42
computer, 93, 104, 105
computing, 10, 71, 113
conductivity, ix, 66, 68
conotoxins, x, 135, 136, 137, 143, 145, 146
conservation, vii, viii, 1, 9, 19, 20, 21, 25, 26, 30, 32, 36, 37, 63, 71, 120, 136, 137, 143
construction, vii, viii, x, 36, 115, 116, 118, 119, 120, 121, 123, 124, 126, 128, 129, 130
consumer surplus, 18
consumers, 9, 127
consumption, x, 8, 9, 115
contaminant, 126
contamination, viii, 36, 38, 39, 44, 53, 54, 57, 59, 60, 61, 123, 125
Conus ventricosus, x, 136, 137, 144, 145, 148

convention, 63
Convention on Biological Diversity, 37, 63
copper, 39
correlation, 110, 111
correlation coefficient, 110
correlations, 101
corrosion, 67
cost, 10, 11, 12, 15, 16, 18, 30
cost-benefit analysis, 16
covering, 117
Croatia, 89
crops, 146
crust, 53, 63
crystal structure, 147
culture, 7
CV, 11, 140
cycles, 43
cycling, 7
cysteine, x, 135, 136, 137, 138, 140, 142, 143, 145
cystine, 146

D

danger, vii, 1, 24
data collection, 41, 51, 60
database, 11, 14
DDT, 46, 59, 61
decision-making process, 26
decomposition, 116
deficiency, 53
degradation, 3, 4, 5, 6, 20, 24, 25, 55, 57, 59, 75, 126, 138, 141, 142, 143
degradation process, 75
Delta, 20
democracy, 23
Department of Agriculture, 38
deposition, 59, 63, 125
deposits, 31, 86
depth, vii, ix, x, 14, 15, 39, 40, 42, 47, 49, 50, 51, 69, 72, 73, 74, 75, 76, 78, 79, 80, 81, 82, 83, 87, 92, 93, 99, 101, 103, 104, 106, 109, 112, 113, 117, 128
derivatives, 55, 118
detection, 44, 45, 46, 47, 127
diffusion, 75, 86, 119
diffusivity, 68
digestion, 44
dipoles, 96
discharges, viii, 36, 59, 66, 67, 71, 88, 112, 132
disinfection, 125
dispersion, ix, 60, 66, 67, 68, 70, 75, 86, 87, 93, 94, 96, 125
displacement, 70
disposable income, 16
dissolved oxygen, 70, 124
distilled water, 43
distribution, 28, 56, 57, 60, 68, 69, 117
diversification, 138
diversity, viii, 36, 103, 120, 128, 129
DNA, 127
DNA damage, 127
DOI, 61, 113
domestic demand, 116
dominance, 57
double counting, 7
drinking water, 22
drug design, 146
drug discovery, x, 136
drugs, 137
dumping, viii, 24, 36, 37

E

early warning, 122, 126
ecological systems, 3, 5, 27
ecology, 29, 133
economic activity, 10
economic indicator, 6, 16
economic systems, 5
economic theory, 30
economic welfare, 17
economics, 26, 28, 29, 32
ecosystem(s), vii, viii, ix, 1, 2, 3, 4, 5, 6, 7, 8, 9, 10, 11, 14, 15, 16, 17, 18, 19, 20, 21, 22, 23, 24, 25, 26, 27, 28, 29, 30, 32, 36, 37, 38, 60, 66, 67, 87, 113, 118, 120, 121, 126, 129, 130
EEA, 5, 29

effluent(s), ix, 66, 67, 68, 70, 71, 72, 73, 75, 78, 81, 82, 83, 84, 85, 87, 88, 124, 132
electricity, x, 115, 116
emission, 62, 97
empirical studies, 11, 14
employment, 127
endocrine, 59, 60
energy, 24, 68, 71, 75, 86, 92, 94, 95, 97
enforcement, 22
environment(s), vii, 1, 2, 3, 4, 5, 14, 16, 20, 23, 24, 36, 59, 67, 92, 94, 97, 102, 120, 125
environmental change, 123
environmental characteristics, 129
environmental conditions, ix, 66, 98, 129
environmental degradation, 24
environmental economics, 16
environmental effects, 126
environmental factors, ix, 68, 92
environmental impact, viii, ix, 36, 66, 129, 130
environmental management, 27
environmental policy, 3
environmental protection, 121, 130
Environmental Protection Agency (EPA), 44, 56, 71, 88, 132, 133
environmental quality, 119, 122
environmental resources, 86
environmental services, 5
environmental standards, 6
environmental threats, vii
equilibrium, 73, 78
equipment, 18, 98, 99, 100, 103, 104
erosion, 19, 20, 29, 101, 117, 118, 129
Europe, 2, 3, 6, 19, 29, 113
European Parliament, 61, 131, 132
European policy, 24
European Union (EU), 26, 28, 29, 37, 59, 61, 102
evaporation, 67
evolution, 5, 67, 130, 147
exploitation, 24, 125
exposure, 126, 127
externalities, 21
extinction, ix, 92, 93, 96, 97, 98, 101, 103, 104, 105, 106, 108, 109, 110, 111, 112, 113
extraction, viii, 36, 37, 44, 45, 46, 66, 67, 86
extracts, 44, 45, 46, 47

F

farms, 120
fauna, 120, 129
fertility, 7
fertilization, 127
field data, ix, 66, 68, 73, 74, 78, 79, 80, 81, 82, 83, 84, 85
filters, 100
financial, 112
financial support, 112
fish, x, 7, 8, 22, 38, 63, 97, 102, 116, 118, 120, 122, 123, 125, 127, 128, 140, 141, 145
fisheries, 8, 14, 21, 24, 27, 37, 38
fishing, viii, 2, 31, 33, 36, 37, 129
fishing activity, viii, 36
fixation, 3, 27
floods, 103
flora, 117, 120
flora and fauna, 117
fluid, ix, 66, 67, 71, 72, 73, 81, 126
fluorescence, 70, 97
fluorimeter, 124
food, 24, 95, 128
food chain, 95
food web, 128
formation, vii, viii, 7, 57, 66, 96, 130
formula, 52, 101, 106
foundations, 32
fragments, 49
France, 62, 148
freshwater, 101
functional analysis, xi, 136
functional approach, 31
fungi, 97
fusion, 145, 148

G

gasification, 128
gel, 47
gene expression, 147
genes, 137
genus, 136
geology, 61
geometrical parameters, 74
GIS, 26, 32
goods and services, vii, 1, 4, 5, 10, 14, 20, 23, 25, 26, 29
governance, vii, 2, 24, 36
government intervention, 10
GPS, 40, 42
grain size, 118, 125
graph, 110
grass, 57
grasslands, 14
gravity, 2, 72
Greece, 32
grids, 42
Gross Domestic Product (GDP), 6, 16, 17
growth, 2, 19, 23, 31, 71, 92, 117, 127
guidance, 117
guidelines, 3, 23, 25, 67
Gulf of Mexico, 67

H

habitat(s), 7, 15, 24, 29, 66, 113, 118, 120, 138, 145
Hawaii, 63
hazards, viii, 36
HDPE, 43
health, viii, 23, 26, 59, 66, 113
heavy metals, viii, 36, 60, 67
height, 47, 73, 81
hexane, 44, 45, 46
highways, 24
historical data, 39
history, vii, 1, 2, 3, 23, 30
homogeneity, 52
human, vii, viii, 1, 2, 3, 4, 5, 8, 9, 14, 17, 24, 25, 36, 37, 56, 97, 98, 111, 121, 126
human development, 2
human health, viii, 36
human welfare, 5, 8, 25
humus, 95
hunting, x, 136, 137, 141, 145, 148
hydrocarbons, viii, 36, 67, 125
hydrogen, 43, 95, 96
hydrogen atoms, 96
hydrogen bonds, 96
hydrogen peroxide, 43
hydrophobicity, 143

I

illumination, 106
image(s), 41, 49, 51, 120
immersion, 104
impact assessment, 86
imports, 116
income, 8, 21
incomplete combustion, 56
individuals, 17, 20, 129
industrial wastes, 38, 56, 71
industry, 21, 22, 68
infancy, 11, 21, 23
insects, xi, 97, 136, 144, 146
institutions, 20, 37
interface, 71, 100
investment(s), 17, 18
ion channels, x, 135, 146
Iran, 132
islands, 2, 138
isolation, 6, 22
isotope, 61
issues, 6
Italy, x, 31, 35, 39, 65, 69, 115, 116, 130, 135

J

Japan, 62
jurisdiction, 37

K

K^+, xi, 136, 140, 141, 147

L

lakes, 71, 98
landscapes, 29
larvae, xi, 136, 145, 146
laws, ix, 61, 91, 102
lead, vii, x, 1, 4, 119, 136
leakage, viii, 36, 38, 59
leucine, 140, 147
light, 12, 92, 93, 94, 95, 96, 97, 98, 100, 101, 105, 106, 111, 112, 113
light beam, 96
lipids, 95
liquefied natural gas, vii
Liquefied Natural Gas (LNG), x, 115
lobbying, 22

M

magnesium, 95
magnitude, 9, 18, 98
mammals, 130
management, 3, 4, 5, 7, 21, 22, 24, 25, 26, 27, 29, 30, 31, 36, 63, 66, 120
marine environment, vii, viii, ix, x, 3, 23, 24, 25, 26, 36, 37, 52, 60, 61, 66, 67, 68, 86, 87, 89, 116, 117, 118, 119, 136
marine species, 117
marketing, 61
marketplace, 23
mass, 45, 46, 71, 100, 138, 139, 143
mass spectrometry, 46, 138
materials, viii, 7, 24, 36, 44, 62, 122
matrix, 122, 123
matter, 3, 55, 57, 92, 94, 95, 96, 119, 128
measurement(s), ix, x, 40, 66, 67, 92, 93, 98, 100, 101, 104, 106, 108, 111, 113, 121
media, 100, 131
median, 141
medical, 60
Mediterranean countries, 2, 19
Mediterranean Sea, 1, iii, v, vii, viii, x, 1, 2, 4, 20, 24, 27, 28, 38, 61, 62, 65, 66, 68, 69, 86, 136, 137, 138, 143, 145
memory, 100, 101, 129
meta-analysis, 15
metals, 52, 53, 54, 62, 63, 118, 119, 125
meter, 18, 20, 42
methodology, 9, 12, 14, 16, 23
microwaves, 44
migration, 2
military, viii, 36, 38
mixing, ix, 55, 66, 67, 68, 71, 72, 73, 75, 82, 83, 85, 86, 119, 120, 128
model system, 145
modelling, 68
models, x, 4, 5, 25, 71, 88, 92, 113, 132, 144
modifications, 120, 121, 126, 128, 131, 136, 141
molecular mass, 54
molecular weight, 138
molecules, x, 95, 96, 135, 136
mollusks, 119, 127, 136
momentum, 71, 72
morphology, 86, 118, 119, 120, 122
mortality, 128
mosaic, 42, 51, 120
motif, x, 135, 136, 146
mRNA, 137
multiplier, 25
mussels, 122, 127
mutagenesis, 148
mutant, 145

N

Na^+, 143, 148
National Research Council, 39
natural gas, ix, x, 66, 86, 115, 116
natural habitats, 138
natural resource management, 24
natural resources, 3, 4, 5, 20
negative effects, 60

Netherlands, 29
neural network, 110, 113
neurons, 110, 141
neuroscience, x, 136
New England, 18, 30
nitrogen, 27, 44, 55, 62, 63, 95
nitrogen compounds, 55
NMR, 141, 142, 148
normal development, 97
North Adriatic Sea, vi, vii, 115
nutrient(s), 55, 70, 92, 102, 118

O

oceans, 3, 37, 93, 98
oil, viii, 36, 37, 39, 65, 66, 67, 68, 71, 86, 89, 131
operating range, 42
operations, ix, 23, 26, 40, 66, 67, 68, 86, 87, 118
opportunities, 22, 23, 32
organic compounds, 59, 60, 119
organic matter, 3, 55, 56, 57, 92, 93, 95, 96, 97, 112, 118, 125
organize, 5
organotin compounds, 125
overlap, 28, 39
oxidation, 100
oxygen, 95, 96

P

Pacific, 88, 141
pain, 148
parallel, 39
paralysis, 145
partition, 59
pathways, 38, 59
PCA, 127
PCBs, viii, 36, 45, 59, 60, 61, 125
peer review, 11, 15
peptide(s), vii, x, 95, 135, 136, 137, 138, 141, 142, 143, 144, 145, 147, 148
peroxidation, 127
Persian Gulf, 116
Persistent Organic Pollutants, 56
personal communication, 19
perylene, 44
PES, 21, 22, 29
pesticide(s), xi, 46, 59, 125, 136, 146
petroleum, 57, 61, 70, 116
pH, 70, 124
pharmaceutical, 146
phenolic compounds, 95
phosphates, 59
photosynthesis, 3, 92, 93, 94, 95, 113
phylum, 129
physical properties, 126
phytoplankton, 55, 92, 94, 95, 96, 102, 112, 113, 128
piano, 88
pipeline, x, 115, 116, 117, 119, 120, 122, 123, 124, 126, 129, 131
plankton, 122
plants, 21, 55, 97, 146
platform, ix, 14, 16, 38, 66, 68, 70, 71, 72, 73, 74, 75, 77, 78, 86, 87
policy, viii, 3, 4, 6, 14, 25, 36, 37, 61, 87, 117
policy makers, 14
policy responses, 6
pollutants, viii, 19, 36, 38, 56, 60, 66, 70, 118, 127
pollution, vii, viii, 36, 37, 39, 53, 54, 71
polycyclic aromatic hydrocarbon, 61, 62
polymerase, 137
polypeptides, 136
ponds, 113
population, 2, 102, 113, 123, 138
portfolio, 23
positive externalities, 21
potassium, 147
predators, 97, 138, 145, 146
preservation, 25, 55, 120
prevention, 7, 87
principles, 133
private investment, 17
private ownership, 16
private sector, 23

probability, 10, 93
probe, 40, 70, 98, 99, 100, 104, 121, 124
produced formation water (PFW), vii, viii, 66
profit, 22
project, x, 26, 28, 112, 115, 116, 117, 119, 122, 124, 130
propagation, 93
property rights, 9
protection, 3, 7, 16, 17, 20, 21, 22, 23, 25, 125, 129
proteins, 55, 95
prototype, 143
public goods, 5, 9, 20, 26
public policy, 25
public service, 5
public-private partnerships, 4, 20, 22
purification, 7
purity, 44
PVC, 100
pyrolysis, 57

Q

quality control, ix, 45, 46, 47, 91
quality standards, 125
quantification, vii, 1, 24, 32, 59
quartz, 52

R

radiation, 43, 93, 94, 95, 97
raw materials, 8
reactions, 95
reagents, 44
receptors, x, 135, 143, 146
recognition, 3, 4, 6, 25, 103, 129, 144, 145
recommendations, viii, 16, 21, 36
reconstruction, 48, 49, 50, 51
recovery, 145
recreation, 17, 24, 30
recreational, 8, 17, 18, 22, 30, 33, 120
reforms, 24
regression, x, 10, 92, 105
regression analysis, 10
regulations, 61, 67, 117, 124
regulatory requirements, 37
remediation, 60
remote sensing, 119, 121, 123
reproduction, 97, 128
requirements, 113, 116, 117, 124
reserves, 28, 37, 68
residuals, 67
residues, 136, 142, 143, 144, 147
resilience, 27
resolution, 42, 47, 49, 50
resources, viii, 7, 9, 10, 18, 24, 29, 36, 37, 60, 118
response, 45, 46, 47, 69, 128
restoration, 4, 22
restrictions, 61
rice field, 102
rights, 30
risk(s), viii, 7, 9, 22, 23, 30, 36, 44, 59, 60, 66, 67, 86, 87, 125
risk assessment, 30, 87
risk factors, viii, 66
room temperature, 43, 47
rotifer, 128
routes, x, 2, 39, 115, 116
routines, 74
rules, 24
runoff, 69
Russia, 116

S

saline water, 67
salinity, 70, 73, 75, 77, 101, 124
salts, 101
saltwater, 16, 101, 109
samplings, 127
scattering, 43, 93, 94, 96, 97
science, 22
scope, 14
SEA, 6
seabed, viii, 36, 41, 50, 60, 66, 116, 118, 119, 120, 122, 124, 126, 129, 130
seafood, 24

Secchi depth, vii, x, 92, 101, 106, 109, 112, 113
Secchi Disk, ix, x, 91, 92, 98, 103, 106, 109, 111, 112
security, 8, 22
sediment(s), 8, 21, 22, 29, 30, 38, 39, 42, 43, 44, 48, 52, 53, 54, 55, 56, 57, 59, 60, 61, 62, 63, 70, 96, 118, 119, 120, 121, 122, 123, 124, 125, 126, 127, 128, 129
sedimentation, 117, 118
sensors, 70, 100, 121, 124
service provider, 21
services, vii, 1, 2, 3, 4, 5, 7, 8, 9, 10, 11, 14, 15, 16, 17, 18, 19, 20, 21, 22, 23, 24, 25, 26, 27, 28, 29, 30, 32, 33
sewage, 57
shape, 51, 96, 97, 126
shear, 71, 73
shellfish, 125
shelter, 118
shipwrecks, vii, viii, 36, 38, 39, 40, 42, 43, 47, 51, 52, 57, 60
shoreline, 8
showing, 10, 69, 80, 110, 112, 142
Sicily island, viii, 36
signal peptide, 137
signals, 61, 126
silica, 44, 47
silver, 44
simulation(s), 73, 75, 79, 80, 81, 82, 83, 87
smoothing, 42
social relations, 8
society, 7, 21, 22, 23, 25, 26
sodium, 101, 117, 124
software, 40, 42, 100
solubility, 59
solution, 45, 46, 47, 93, 141, 148
sound speed, 40
Spain, 1, 16, 17, 26, 27, 32, 33, 91, 117
speciation, 125
species, x, 5, 9, 19, 29, 94, 95, 97, 118, 119, 122, 126, 127, 128, 129, 130, 135, 137, 138, 141, 145
spectroscopy, 141, 142
SSS, 51
stability, 23, 45, 46, 47, 72, 80, 81, 83, 87, 127
stabilization, 8
stakeholders, 21, 22, 25
state, 98, 101, 111, 121
steel, 100
storage, 98, 99, 100, 142
storms, 103
strategy use, 118
stratification, 68, 73, 77, 81, 83, 86, 87, 109, 110, 112, 124
stress, 127, 130
stressors, 4
structural units, vii, 1, 5, 24
structure, x, 24, 25, 70, 94, 96, 116, 118, 119, 123, 124, 125, 128, 129, 130, 136, 141, 142, 143, 146, 147
subjectivity, ix, 92, 98, 111
substitution(s), 136, 137, 140
substrate(s), 117, 118, 120, 130
surface layer, 81
sustainability, 22, 23, 27, 28, 29
sustainable development, 23, 33
symptoms, 145
synthesis, 144

T

target, xi, 39, 41, 42, 120, 126, 136, 137, 140, 142, 143, 144, 145, 146
target identification, 41, 42
taxa, 127, 128, 129
taxes, 10, 18
techniques, 9, 10, 11, 12, 14, 18, 25
technology, 37
temperature, ix, 45, 46, 57, 66, 68, 70, 73, 75, 77, 95, 99, 100, 117, 118, 124, 128
temporal variation, 55
terminals, 131
territorial, 30
thermodynamic properties, 56, 57
threats, 28
tides, 92
time series, 73
tobacco, 146

tourism, 2, 14, 17, 22, 23, 24
toxic effect, 68, 127, 128
toxic material, viii, 36
toxicity, 59, 127, 128, 145
toxin, 143, 144, 145, 146, 147
trace elements, 61, 125
trade, 2
trajectory, 71, 73, 82, 83, 94, 96
transactions, 21
transformation, 3, 16, 17
transparency, 96, 98, 102, 103, 111, 112
transport, ix, 8, 22, 24, 59, 66, 67, 70, 93
transportation, 125
treatment, 62, 104, 148
tryptophan, 140, 147
turbidity, ix, 92, 101, 118, 119, 124
turbulence, 73, 75
Turkey, 30

U

UK, 13, 26, 88
UNESCO, 21
unions, 96
United Nations (UN), 3, 22, 37, 63, 132
urban, 2, 14, 55, 56, 63
urban areas, 2
USA, 74
USGS, 63
UV, 101, 108, 111

V

Valencia, 91, 98, 102, 109, 110
valuation, vii, 1, 2, 6, 7, 8, 9, 10, 11, 12, 14, 15, 16, 17, 18, 19, 22, 24, 25, 26, 27, 29, 30, 31, 32, 33
variables, 119, 121
variations, 55, 77, 124
vector, 15
vegetation, 101, 113, 117
velocity, 40, 71, 72, 73, 74, 77, 87
venom, vii, x, 135, 136, 137, 138, 141, 142, 143, 145, 147, 148
ventricosus, vii, x, 136, 138, 139, 140, 141, 142, 145, 146
vertebrates, 145
vessels, viii, 36, 38, 40, 44, 60, 119
vision, 3, 25
vitamins, 95
volatilization, 67

W

Washington, 28, 31, 132
waste, viii, 36, 66, 67
wastewater, 55, 59
water bodies, ix, 92, 93, 95, 98, 101, 103, 111, 112, 113
water quality, vii, ix, 91, 93, 111, 113, 133
watershed, 103
wavelengths, 94, 101, 105, 106, 108, 109, 112
welfare, 3, 4, 7, 19, 25, 26
well-being, vii, 1, 3, 5, 8, 24
wetlands, 14, 16, 30, 101, 102, 120
wild type, 146
wildlife, 59
wood, 57
worldwide, ix, 2, 21, 24, 66, 67, 71, 91
worms, 145

X

X-ray diffraction (XRD), 43

Y

yield, ix, xi, 3, 9, 18, 66, 67, 136

Z

zooplankton, 55, 122, 128